The Macat Library
世界思想宝库钥匙丛书

解析约翰·斯图亚特·穆勒
《功利主义》

AN ANALYSIS OF
JOHN STUART MILL'S
UTILITARIANISM

Patrick Tom　Sander Werkhoven ◎ 著

陈琦 ◎ 译

上海外语教育出版社
外教社 SHANGHAI FOREIGN LANGUAGE EDUCATION PRESS

目　录

CONTENTS

引言

要 点

- 约翰·斯图亚特·穆勒（1806—1873），英国哲学家、政治活动家、议会议员，他在著作和政治生涯中捍卫个人自由，发展了功利主义的道德哲学。

- 《功利主义》（1861）认为幸福是人生中最大的善，由道德上的良好行为产生。

- 《功利主义》作为道德哲学（哲学中讨论伦理的一个分支）的一部重要著作，清晰地解释了功利原则*，并为之辩护。

约翰·斯图亚特·穆勒其人

《功利主义》（1861）的作者约翰·斯图亚特·穆勒是英国哲学家、经济学家、公务员。他出生于1806年，是苏格兰哲学家、历史学家詹姆士·穆勒*与他的妻子哈丽特·伯罗的儿子。詹姆士·穆勒从小教儿子古代哲学和功利主义的道德理论，这方面他得到了哲学家与法学家*（法律学者）杰里米·边沁*的帮助，边沁通常被认为是功利主义的奠基人。两人都希望小约翰·斯图亚特·穆勒将来能成为功利主义的信奉者和传承人。穆勒实现了他们的期望。

穆勒作为一位政治活动家和重要的社会改革者，毕生写下许多文章和哲学著作。在他发表的第一部重要著作《逻辑学体系》（1843）里，他支持逻辑可以作为一种证明方法。1848年，他出版了《政治经济学原理》，这是19世纪经济理论的扛鼎之作。他在1859年的《论自由》一书中捍卫了个人自由并论证对他人造成伤害的行为应受限制。1863年，他出版了《功利主义》，在书中阐释并捍卫了他的导师杰里米·边沁的道德理论。

　　穆勒活跃于英国重要的社会变革时期，在当时就已广为人知。他担任了三年的自由党议会议员，该政党以个人主义信念为政策基石。在此期间，他发起予以妇女投票权的运动。穆勒与哲学家和活动家哈丽特·泰勒 * 在经历 20 年的友谊后于 1851 年结为连理。泰勒本身就是位伟大的思想家，她帮助穆勒发展了哲学、经济学和政治学等理论。

《功利主义》的主要内容

　　《功利主义》讨论道德哲学，尤其是人类应该如何行动以及如何生活的问题。穆勒在书中论证，幸福是人生的目标，只有增进人类幸福的事物才能被认为是"善"的。穆勒认为这一基本论点可以得到经验主义 * 的证明（即从可观察的证据出发进行推理证明），他指出，人类的确实际上都试图获得幸福。他将"幸福"定义为快乐的感受和痛苦与苦难的解除（该定义可以被理解为"快乐主义的" *）。因此人生的目标在于体验快乐、避免痛苦与苦难；这也就是至善。

　　尽管他的观点取自导师兼前辈杰里米·边沁，穆勒指出该观点在古希腊，特别是哲学家伊壁鸠鲁 * 的伦理学著作中也出现过。他发展了这一传统观点，进一步将快乐区分为高级快乐和低级快乐，承认并非所有的快乐在衡量幸福方面都同等相关。在评估幸福时，他主张高级的快乐更为重要，比如那些我们从诗歌、音乐以及洞察力的发展中获得的快乐。

　　从幸福是人生的终极目标这一信念出发，穆勒推导出书中最重要的论点——即一个行为如果增进幸福并减少苦难，那么它在道德上就是善的；反之，在道德上就是恶的。重要的是，道德上的善行所增进的并不是个人的幸福，而是受到此人行为影响的所有人幸福的总和。穆勒写道，"道德之根本，也即功利，或最大幸福原则*……认为行为的对错与它们促进幸福或导致不幸的倾向正相关"[1]。简而言之，这就是功利原则。

虽然上述道德理论暗示了个体的幸福可能不得不被牺牲，穆勒还是在《功利主义》中为之进行了辩护，并对同时代人提出的对功利原则的诸多批评作出回应。在这些批评声中，功利原则与正义感相矛盾的反对意见尤为引人注目。在书的最后部分，穆勒讨论了正义与功利的关系，认为正义也是以功利原则为基础的，因此，正义和道德正确之间并不存在冲突。

《功利主义》的学术价值

《功利主义》在哲学史上是一部重要著作。它对功利主义系统而细致的辩护使得该原则成为西方哲学三大道德理论之一。

穆勒非常重视良好的道德教育。他的著作面向大众，旨在教导人们如何思考道德问题，以使自身变得更美好、更幸福。人们都想知道在某些特定情境下该如何行动，因而经常自问，"人世间做什么才是正确的？"穆勒的功利主义给了我们一种答案。他写道，正确的事情就是可以带来最大幸福和最少不幸的事情。这是比其他大多数道德理论更实用的原则，穆勒对这个观点的辩护很可能引导许多读者对何为正确的行为方式产生了不同的思考。

然而，《功利主义》绝不仅是一部教育性的著作；作为功利主义的奠基性文献，它清晰定义并捍卫了功利原则。在穆勒所处的时代，许多人认为功利主义作为一种道德哲学理论是有缺陷的。但是穆勒充分、全面、高超的辩护使得功利主义成为 20 世纪道德哲学中最重要的理论之一。尽管自穆勒时代以来，功利主义已经发展出许多不同的流派，但他们都以穆勒的观点为基础，并且用《功利主义》来支持论点。

因此，《功利主义》不仅是道德哲学最重要理论之一的奠基之作，它在我们应该如何行动以及生活的讨论中，也一直具有很强的现实意义。

---

1. 约翰·斯图亚特·穆勒，罗吉尔·克里斯普编：《功利主义》，牛津：牛津大学出版社，1998 年，第 55 页。

第一部分：学术渊源

1 作者生平与历史背景

要点 ✐—

- 穆勒的《功利主义》在道德哲学（对伦理的哲学思考）历史上是一部奠基之作。它将英国社会改革家、哲学家杰里米·边沁的功利主义观点发展为精致全面的道德理论。
- 穆勒从小就被培养成为功利主义的有力推动者。杰里米·边沁在他的教育中亲自扮演了重要角色。
- 穆勒写作时正处于社会和政治变革时期，他通过支持自由主义 *（为促进和捍卫个人自由而组织的社会运动）来回应变革。

为何要读这部著作？

《功利主义》（1861）的作者约翰·斯图亚特·穆勒是 19 世纪最有影响力的思想家和自由主义改革家之一。在他最为广泛阅读的哲学经典中，这部著作定义并从哲学层面论证了功利主义这一伦理理论 *，修正发展了先前功利主义理论家的成果。

"伦理理论"是对人类正当举止、行动的合理解释；根据功利主义原则，我们应该促进善（这里的善意味着总体上的幸福多于不幸）。尽管这个理论与穆勒和英国哲学家、法学家杰里米·边沁相关联，其核心思想却并非起源于他们。

穆勒的《功利主义》对功利主义进行了相当完善的阐释，至今仍是道德哲学最重要的文本之一。在这本书中，穆勒区分了高级快乐——例如诗歌和洞察力的发展——以及低级快乐，认为前者在幸福的总和中更为重要。在功利主义信仰体系里，他把人类的同情心放在核心位置，而不是他的老师边沁提出的"最大幸福原则"（后者本质上是一个更精于计算*——即以计算结果为准的观念）。穆勒也详细论证了我们为何要根据功利主义道德行事。

穆勒试图通过这些对功利主义的修正、发展来为之辩护，对抗至今仍见于相关文献的一些反对意见。因此，人们对穆勒《功利主义》的兴趣未见消减不仅是因为它在道德哲学上的奠基角色；也因为它对今日尚存的反对功利主义的声音提供了有效回应。

> "在（我父亲）写作的同一个房间和同一张桌子上，我预习了全部希腊语课程……我被迫就每个我不知道的单词求助于他。他，最没有耐心的男人之一，却甘受这种无休止的打扰，并在这种打扰下撰写了数卷《历史》以及那些年里他必须写的所有其他东西。"
>
> ——约翰·斯图亚特·穆勒：《自传》

作者生平

穆勒 1806 年出生于伦敦，1873 年去世。他的父亲詹姆士·穆勒赞同 17 世纪英国哲学家约翰·洛克*的学说。洛克宣称人的心灵在出生时是一块"白板"*，而知识通过经验获得。詹姆士·穆勒希望在自己和朋友杰里米·边沁过世后，儿子能继续功利主义的政治与社会改革。他开始着手严格的教育计划，亲自辅导儿子，以求造就一个完美的功利主义头脑。[1] 在自传中，约翰·斯图亚特·穆勒记录了父亲如何敦促他学习希腊语和拉丁语，并要求他系统性地浏览所有的古典名著。

穆勒在自传里记录他被要求学习古希腊语，就在他父亲工作的"同一个房间和同一张桌子上"。他写道，"我被迫就每个我不知道的单词求助于他。"[2]

但是，这样的教育并未立刻产生期待的效果。1826年，穆勒遭受了严重的精神危机。他的最终康复得益于与诗歌的接触，特别是英国诗人威廉·华兹华斯*和萨缪尔·泰勒·柯勒律治*的作品，这些作品源于感情的抒发，具有典型的浪漫主义*传统。穆勒希望通过阅读这些诗歌来提高自己的美学情趣。[3]

康复之后，他开始形成自己的哲学观点。他强调情绪和感情在人类生活中的重要性，如此一来他的思想便与在某些方面更精于计算的边沁功利主义区别开来。[4]不过，穆勒至死都对功利主义和经验主义（认为知识应当以经验为基础的一种信念）坚信不疑。在自传中，穆勒赞扬妻子哈丽特·泰勒·穆勒帮助自己发展了哲学思想。

穆勒还是位政治活动家，1865年到1868年作为自由党成员供职于议会。他是第一位支持妇女拥有投票权的英国议会议员，并且终身积极拥护该目标。

创作背景

在穆勒生活的时期，英国正从以农业和手工劳动为基础的经济向以工业生产为基础的经济转型，由此带来了社会和经济的巨大变化。这一时期被称为"工业革命"*。它促进了英国的经济增长，但也带来了许多挑战，包括贫富差距加大、城市化迅速发展、失业增加、犯罪频发、物价上涨、贫困人口大幅上升、工资下降、住房恶化、疾病激增。19世纪政治改革活动及辩论占据主导地位（这在一定程度上是受1789年到1799年法国大革命*的影响，大革命期间法国国王被处决，政府也彻底改革）。人们强烈要求个人自由、民主和议会改革、最少干预的政府、自由企业经济，渴求建立更为宽容、公平的社会。这一时期也有人呼吁用代议制政府取代贵族统治。在 19 世纪的英国，自由主义政治运动者反对贵族统治，他们相信议会由一群腐败的精英组成，他们为了自己的利益行事，侵犯个人自由。

总之，穆勒生活在一个社会动荡的时期。这些挑战为他的许多作品和哲学活动带来启发。他对不断变化的环境作出回应，为自由主义摇旗呐喊。

---

1. 约翰·斯图亚特·穆勒，杰克·斯特灵格编：《自传》，伦敦：牛津大学出版社，1971 年。
2. 穆勒：《自传》，第 9 页。
3. 参见穆勒：《自传》；托马斯·伍兹：《诗歌与哲学：约翰·斯图亚特·穆勒思想研究》，伦敦：哈奇森出版社，1961 年。
4. 参见约翰·斯图亚特·穆勒："边沁"，阿兰·瑞安编，《功利主义及其他论文》，伦敦：企鹅图书，2004 年。

2 学术背景

要点 ⌐━━

- 《功利主义》是一部道德哲学著作，为"人应该怎样生活"的问题提供了一种答案。

- 在书中，穆勒提供了一种有别于古典德性伦理学*（一种用于理解"正确"行为的方法，可追溯到古希腊，侧重于个人美德行为的修养）和道义论*（一种基于义务和道德责任*的伦理学方法，主要归功于18世纪德国哲学家伊曼努尔·康德*）的理论。

- 尽管穆勒深受自己的父亲和功利主义哲学家杰里米·边沁的影响，他也从其他古代、现代的哲学家，以及与他同时代的功利主义批评者的观念中，获得了感悟。

著作语境

约翰·斯图亚特·穆勒的《功利主义》是道德哲学领域的一部著作，该哲学领域关注的问题是"人应该怎样生活"。

此问题由古希腊哲学家苏格拉底*提出，可以被进一步分为三个问题："什么是幸福"，"我们如何知道什么是道德上的正确"，以及"我们何来动力去做道德上正确的事情"。

道德哲学关注的不是世上的事物如何，而是事物应该如何。道德哲学不描述人们如何做出决定并处理他们的生活或分析社会如何运作，而是拷问人们应该如何度过他们的人生，社会应该被如何组织。长期以来，人们一直怀疑找到道德哲学问题的答案的可能性。许多人相信如何度过人生应该交由个人来决定。然而，大多数道德哲学家主张，行为有对错，人分道德好坏，社会组织也或多或少有公正、不公之分。穆勒的《功利主义》论证了这样的主张。

> "在当前的著作里，我将不对其他理论作过多讨论，而是试着为理解和认同功利理论或幸福理论提供证据、作出贡献。"
>
> ——约翰·斯图亚特·穆勒：《功利主义》

学科概览

一般来说，道德哲学有三大传统分支：德性伦理学、康德道义论*、功利主义。

德性伦理学通常与古希腊哲学家亚里士多德*的著作联系在一起，亚里士多德是举世闻名的思想家柏拉图*的学生。[1]在亚里士多德看来，决定人类本质的是思考能力。要过上好的生活，我们必须善用思考能力（也就是依照"美德"来思考）。他认为，只有"美德行为者"（换句话说，就是一个充分运用自身理性思维能力的人）才知道在具体情况下该如何行动。德性伦理的观点不是用来决定是非的一般性原则，而是认为受过良好教育和有道德的行动主体会可靠地选择最佳行动方案。

鼎鼎有名的德国哲学家伊曼努尔·康德宣扬的康德道义论则与之相反，审视的不是个人的性格特征，而是行为本身的对错。[2]康德试图建立一个基于"纯粹理性"的普遍性原则，而对是什么让人类快乐不加关注。康德的观点是，道德上的良好行为在于按照一个人能够理性地愿意（即理性地想要某事发生）将其作为普遍法则的原则行事。

因此，如果一个人能够理性地运用意志将一个行动原则作为普遍法则，那么该行为在道德上就是好的；反之（即结果自相矛盾），那么该行为在道德上就是错的。这种道德考验并非想让大家生活在一个所有人都依照特定原则行动的世界里。康德志不在此。关键在于这个原则必须在理性上有可能成为普遍法则。如果它的结果是矛盾的，那么任何建立在该原则上的行为在道德上都是错误的。

然而，根据《功利主义》，行为的善并不取决于个人品格（如德性伦理所说的），也不取决于理性地运用意志使其成为一条普遍法则的可能性（如康德道义论所说的）。功利主义认为行为的善取决于它的结果。如果一个行为产生好的结果，那么这个行为在道德上就是好的，无论行为主体是否为有德之人。功利主义者只在乎一个行为带来的总体的幸福与不幸。在功利主义者看来，道德上正确的行为方式，是带来最大的善：它必须增加世界上的幸福的总和，并减少不幸的总和。

学术渊源

最直接影响穆勒思想的是他的父亲詹姆士·穆勒和他的导师——现代功利主义"奠基人"杰里米·边沁。两人都属于包括英国经济理论学家大卫·李嘉图*等人在内的"哲学激进主义者"*的思想流派。但是穆勒青年时代的思想世界"结合了杰里米·边沁式的激进主义的开放性……和他通过接触古代思想所产生的小心翼翼，即并非所有的关于善的观点都可以被简化为总体的快乐"[3]。

因此，穆勒的思想也受到了古希腊哲学的影响。例如，幸福是人生的最终目标这一观点，也在诸如亚里士多德、伊壁鸠鲁、阿里斯提波*以及柏拉图等古代哲学家的著作中出现。功利的教条早已被在穆勒之前和与他同时期的思想家们所接受。例如，克劳德·阿德里安·爱尔维修*、切萨雷·贝卡里亚*、约翰·洛克、大卫·休谟*以及弗兰西斯·哈奇森*等都从中汲取思想源泉。

功利主义的反对者包括苏格兰哲学家、数学家和社会评价家托马斯·卡莱尔*、英格兰思想家约翰·罗斯金*等。这些批评者并不把功利主义看作是"道德的基本原则和道德责任的来源"[4]。他们相信功利主义与真正的道德上的正确性相冲突。卡莱尔甚至宣称，功利主义因它对快乐的唯一关注，事实上是一种"猪的哲学"。穆勒的论证受到了这些批判性声音的影响。

在《功利主义》一书中，穆勒综合了父亲詹姆士·穆勒和导师边沁的思想、古代哲学流派的深刻观点，以及功利主义思想反对者的论证。

---

1. 参见亚里士多德：《尼各马可伦理学》，收录于《亚里士多德全集：牛津版修订译本》，乔纳森·巴恩斯编，新泽西州普林斯顿：普林斯顿大学出版社，1995 年。

2. 关于康德的道德哲学，参见伊曼努尔·康德：《道德形而上学基础》，玛丽·格雷格编译，剑桥：剑桥大学出版社，（1785 年）1998 年。

3. 罗伯特·德维尼：《改革自由主义：约翰·斯图亚特·穆勒对古代、宗教、自由和浪漫道德的使用》，康涅狄格州纽黑文：耶鲁大学出版社，2006 年，第 2 页。

4. 约翰·斯图亚特·穆勒，罗吉尔·克里斯普编：《功利主义》，牛津：牛津大学出版社，1998 年，第 51 页。

3 主导命题

要点 🔑

- 穆勒的《功利主义》试图通过维护和拓展哲学家杰里米·边沁的功利主义哲学来回答关于如何生活和行动的基本道德问题。

- 穆勒与功利主义道德哲学的所有流派交锋——特别是那些反对边沁观点的人们。

- 穆勒仔细探究了功利主义的主张,如何反驳批评者,以及为什么与其他更多地建立在直觉之上的伦理学流派相比,功利主义可以更好地得到经验*证据的支持。

核心问题

约翰·斯图亚特·穆勒的《功利主义》旨在回答"什么是行为或生活的正确方式"。

为了回答这个问题,穆勒尝试解决一系列更深层的道德哲学的中心问题。他首先发问,什么是好的行为的"结果"或"目标"。每当我们认为某事是好的,那是因为我们认定它所导向的结果也是有益的。例如,在学校获得高分是好的,因为它可能帮我们进入研究生院,而进入研究生院是好的,因为它有助于一个人智力的发展,而一个人智力的发展是好的,因为某个更进一步的原因——诸如此类。

在某一时刻,这样的环环相扣不得不停止,一定存在一个最终的善(也就是说存在这样一种事物,一切对它有贡献的东西都是好的)。如果没有最终的善,我们将无法知晓是什么让其他事物成为好的。穆勒认为为了知道如何去行动,我们应该问这样一个问题,

"最终的善由什么构成？"

穆勒对上述问题的回答是幸福。但是这接下来产生更多的问题。幸福是由什么组成的？幸福和道德之间的关系是什么？如果最终的善是幸福，我们在道德上有何义务？如果幸福是最终的善，幸福和正义间的关系是什么？这些问题是穆勒在《功利主义》里所关注的，也是他尝试去回答的。

穆勒为关于如何行动和生活的哲学辩论作出了贡献。这正是古希腊思想家柏拉图、亚里士多德和伊壁鸠鲁等哲学家以及18世纪德国哲学家伊曼努尔·康德等所努力解决的共同争论。穆勒作品中的核心问题与更早期的哲学探究并无二致。

> "从哲学曙光乍现开始，有关'至善'的问题，或道德的基础是什么的问题，构成了思辨思维的主要议题，占据了最具天赋的智者们的头脑，并将他们分裂成不同的宗派和流派，彼此展开激烈的论战。而两千多年后，同样的讨论仍在继续。"
>
> ——约翰·斯图亚特·穆勒：《功利主义》

参与者

在《功利主义》的第一章，穆勒认为归纳伦理学派 * 和直觉伦理学派 * 的追随者们在道德基于原则的确立这一点上是一致的。因此，在写《功利主义》这本书时，穆勒脑海里的主要参与方是这两个学派的辩护者。尽管两个学派都赞同道德需要原则，他们对道德责任（也就是一个人有道德义务去采取的行动）的终极来源存在分歧。在杰里米·边沁版本的功利主义里，快乐和痛苦是支配人类行为的唯一事物，而功利原则（或最大幸福原则）是道德的基础。[1]
直觉学派的追随者则不同意这一观点。

穆勒也试图将他父亲詹姆士·穆勒和杰里米·边沁的思想与
其他思想家结合起来，包括托马斯·卡莱尔（与浪漫主义文学、艺
术、哲学运动相联系的苏格兰哲学家）、古代哲学家如苏格拉底和
伊壁鸠鲁等，以及那些赞同实证主义*信条的人们，他们相信科学
知识必须从感官经验中获得并通过实验加以检验，例如法国社会理
论家奥古斯特·孔德*。在穆勒的著作中，孔德实证主义的影响清
晰可见，因为穆勒将自然科学视作获得知识的唯一途径。穆勒否定
德国哲学家伊曼努尔·康德和英国神学家与哲学家威廉·休厄尔*
等思想家们的观点，即存在先验*伦理原则——也就是独立于经验
之外而推断出来的伦理原则。

当代论战

尽管同时代的辩论受到归纳和直觉伦理学学派支持者们的影
响，穆勒认为人们对于功利主义本身也还有着诸多误解。除了争论
功利主义是不是正确的道德理论，还有许多人讨论何为功利主义的
首要主张。穆勒在书中直面那些认为社会利益增加并不能促进幸福
的人们。他指出，他们所否认的正是功利主义所秉持的。

穆勒同时期的另一论战是功利主义是否就如托马斯·卡莱尔
所形容的，是一种"猪的哲学"。卡莱尔的批评主要针对边沁的功
利主义。边沁相信，功利（或幸福）可以通过把人们将会经历的所
有快乐加在一起来测量。批评者们认为，这种形式的功利主义是一
种动物般的道德理论，因为它促进的是一种支配动物行为的粗俗的
快乐。穆勒为功利主义辩护，主张（反对边沁）快乐的质量至关重

要，一些快乐比其他快乐质量更高——因而具有更高的价值。这与边沁的主张相左，边沁认为在强度、持续时间和确定性等方面具有相等性的快乐也必然具有相同的价值。

穆勒还批评了伦理直觉主义*—— 一种认为我们拥有道德真理的先验知识的信条——因为他认为这仅仅是对现状的反映。他在经验主义的基础上为功利主义辩护，宣称经验告诉我们，人们确实试图以促进自身幸福的方式行动，尤其是由"更高级"的快乐所带来的幸福。

---

1. 杰里米·边沁，J. H. 彭斯和 H. L. A. 亨特编：《道德与立法原理导论》，伦敦：阿斯隆出版社，1970 年。

4 作者贡献

要点 🔑

- 在《功利主义》里，穆勒尝试定义功利主义哲学，并为之辩护，反对若干批评，包括认为功利并不意味着快乐和幸福的批评。

- 《功利主义》采取道德哲学的经验主义方法：其中，穆勒宣称我们可以通过观察自身对幸福的清晰渴望，从而推断出幸福是最大的善。

- 穆勒维护哲学家杰里米·边沁提出的最大幸福原则，但增加了快乐的质量作为衡量幸福的维度。

作者目标

约翰·斯图亚特·穆勒在《功利主义》的开篇指出，功利主义的批评者严重误解了它的中心信条。他继续列举许多反对功利主义的批评，包括：幸福实际上无法获得；如果按照"最大幸福原则"行事，每个人最终都不会满意；功利主义理论仅仅依靠冰冷的计算；总体的幸福过于复杂无法计算；它必然导致相互矛盾的责任；它在道德上没有给上帝留下位置；它只是作为个人行为的一种自我辩护；功利主义是一种相对人类而言更适合动物的道德理论；以及它与关于正义的普遍观念相悖。

但是穆勒在《功利主义》第二章提到的首个批评是指责功利主义集中关注功利而非快乐。穆勒对该误解如此回应："了解此事的人都知道，每一个主张'功利'理论的作者，从伊壁鸠鲁到边沁，所言的'功利'，不是什么有别于快乐的东西，而正是快乐本身，连同免除痛苦在内。"[1]

厘清功利主义的意思之后，穆勒试图为此提供一个经验主义的辩护。他反驳了包括伦理直觉主义在内的其他伦理观点，并主张道德建立在经验和观察的基础上。穆勒与其他功利主义者不同：他认为可以凭实证依据（也就是通过基于可观察性证据的推断）来捍卫功利主义。他对功利主义信条本身的修正也让他与早期功利主义思想家有很大的区别。

> "了解此事的人都知道，每一个主张'功利'理论的作者，从伊壁鸠鲁到边沁，所言的'功利'，不是什么有别于快乐的东西，而正是快乐本身，连同免除痛苦在内；他们没有把有用性与舒适性和装饰性对立起来，而是一直表明'有用性'意味着包括这些（舒适性和装饰性）在内。"
>
> ——约翰·斯图亚特·穆勒：《功利主义》

研究方法

在《功利主义》的第一章，穆勒首先驳斥其他替代性理论，特别是那些依据伦理直觉主义的理论；在第二章，他定义功利主义并将上述反对意见考虑在内；在第三章，他思考人们依从功利主义原则的各种可能动机；在第四章，他提出了他对功利主义的证明；在第五章，他集中关注正义以及它与功利的关系。穆勒并不是简单地驳斥对手来为功利主义辩护，而是完善和发展詹姆士·穆勒与杰里米·边沁对它的阐释。最终，他发展出了独树一帜的功利主义版本。

《功利主义》的原创性在于它的中心主张与经验主义的证明。经验主义的证明让他的著作与法国社会理论家奥古斯特·孔德的实证主义相一致；根据实证主义，科学知识必须以感觉经验为基础并通过实验来验证。穆勒认为，功利主义的经验主义证据就是所有人都渴望自身幸福：

"没有理由可以解释为什么公众幸福值得向往，除非说每个人都在能够获得的范围内渴求自己的幸福。然而，这却是事实，因而我们不仅有被认可的所有证据，而且有一切可能需要的证据。"[2]

因此，穆勒除了创造了他自己的功利主义版本并在异议中为之辩护外，还提供了这则信条的经验主义证据。这一点别出心裁，不落窠臼，也驳斥了那些相信道德直觉是道德来源的哲学家们。

时代贡献

穆勒的导师杰里米·边沁，在其所著的《道德与立法原理》（1789）里宣称，人类生而寻求快乐、躲避痛苦，他们为了快乐本身而追求快乐。他说，快乐和痛苦有身体上的、政治上的、宗教上的和道德上的约束依据。然而，政治的、宗教的和道德的依据，与身体上的依据相比，居于次要位置。最大的幸福只能通过最大化快乐和最小化痛苦得到实现。因此，一个法案或一条法律如果能对牵涉其中的所有人带来最多的快乐和最少的痛苦，那么这个法案或这条法律就是好的。边沁发展了一种可以用来量化快乐和痛苦的科学理性的方法。后来他将其应用到英国法律系统和机构中。

尽管穆勒没有放弃边沁的基本的（功利）原理，但他批评了他的前辈们，包括他的父亲，因为他们没有提出关于善的适当的概念（例如，总体上的幸福多于不幸）。由此，穆勒修正了功利主义并将定性的快乐囊括进功利的概念。这是与他前辈们的功利主义的分野。

这个分野导致了穆勒是个自相矛盾的功利主义者的看法。然而，他在《功利主义》一书中的目的和意图清楚地展现并构成了一个条理连贯、论证完善的方案，尽管关于他对功利原则的辩护一直存在多种解读。例如，当他声称从"个人的幸福对那个人而言是善"的观点出发，可得出公众幸福是"对所有人而言的善"[3]的时候，他被指责在核心论证里犯了逻辑错误。批评者指责他犯了合成谬误*（即一种建立在"对部分而言是正确的事情对总体而言也是正确"的错误推断之上的论证）。研究穆勒《功利主义》的现代学者们，例如哲学学者亨利·韦斯特*、大卫·里昂*、乔纳森·里雷*以及约翰·斯科鲁普斯基*，提供了对于穆勒论证的其他解读，以求揭示他的真正意图。

---

1. 约翰·斯图亚特·穆勒，罗吉尔·克里斯普编：《功利主义》，牛津：牛津大学出版社，1998 年，第 54 页。
2. 穆勒：《功利主义》，第 81 页。
3. 穆勒：《功利主义》，第 81 页。

第二部分：学术思想

5 思想主脉

要点 &-⊶

- 穆勒主张，道德的基础在于一个行为如果提升了总体的幸福，那么它在道德上就是正确的，反之，它在道德上则是错误的。

- 穆勒说，衡量一个行为是否提升了总体幸福，我们不仅应该看它产生的快乐的量——我们也应该考虑由此提升的快乐的质。

- 尽管《功利主义》写得明白易懂，但关于部分章节的确切含义仍有争论。

核心主题

约翰·斯图亚特·穆勒的《功利主义》的核心主题是功利原则、高级和低级快乐，以及良好的道德品质的发展。这本书的中心论点被清楚地陈述为"作为道德基础的原则，功利或最大幸福原则认为，当行为倾向于促进幸福时，这些行为相应就是正确的，当倾向于产生幸福的反面时，它们就是错误的"。[1] 在捍卫这一主张时，穆勒赞同将其导师杰里米·边沁的最大幸福原则作为道德的基础。[2]

人生的终极目标是幸福。人们所渴求的每种事物之所以被渴求，是因为它导向幸福或者提供某种幸福。因此，幸福必定是最终的善。穆勒将幸福定义为"快乐如期而至与痛苦的消失"。他将不幸定义为"痛苦、缺少快乐"。[3] 穆勒主张，不仅是一个人的行为应该被导向最大化幸福，一个人的品格也应该旨在达到这一目标。虽然穆勒不赞同亚里士多德的品格决定正确行为的信条，但他的确认为拥有善良品格是重要的，因为这多半将使个体以促进世界上的幸福的方式行事。

穆勒接着超越了边沁，区分了社会上被体验的快乐的量和质。这是对功利主义批评者如苏格兰思想家托马斯·卡莱尔的一种回应。卡莱尔写道，"假设生活……没有比快乐更高的目的，没有更好的和更高贵的渴求追寻的对象，那就意味着十足的偏狭和卑鄙，是只适用于猪的信条。"[4] 穆勒发展了这个至关重要的观点，除了量以外，快乐的质在测量和排序各种快乐时是重要的。穆勒识别了快乐的两种类型：高级快乐（精神的愉悦），例如诗歌和知识；低级快乐（动物性的快乐），例如食色。为了比较哪种快乐在质上更优越，需要找那些经历过两者的人们核实。穆勒相信，那些经历了高级快乐的个体更喜欢高级快乐而不是低级快乐，并且不会愿意放弃它们，即便是为了"一个可以最大限度地满足野兽快乐的许诺"。[5]对穆勒而言，较高层次的快乐在质上更优越，并且与较低层次的快乐相比更令人向往。

> "作为道德基础的原则，功利或最大幸福原则认为，当行为倾向于促进幸福时，这些行为相应就是正确的，当倾向于产生幸福的反面时，它们就是错误的。幸福为快乐如期而至与痛苦的消失；不幸为痛苦、缺少快乐。"
>
> ——约翰·斯图亚特·穆勒：《功利主义》

思想探究

把幸福定义为人生的终点（也就是目标）后，穆勒并没有得出这样的结论，即我们应该仅仅试图增加我们自己的幸福。相反，他认为我们应该促进整体的幸福——亦即整个社会所经历的幸福的总和。这暗示着在一个不是人人幸福的不完美的世界里，我们有时应当为了他人的幸福而牺牲我们自己的幸福。穆勒认为"作出这种牺牲的意愿是在人类身上可以找到的最高美德"。[6]但是与其他道德理论不同，穆勒补充道这种自我牺牲并不好也并不崇高，"不增加或不倾向于增加幸福总和的牺牲，在功利主义看来是一种浪费。"[7]功利主义宣称道德上正确的行为必须促进普遍的善，而不只是某人自己的。穆勒因此将功利主义与基督教教义"己所欲，施于人；爱邻如己"相比较。在穆勒看来，这"构成了功利主义的完美理想"。[8]

穆勒的功利主义所区分的更高层次和更低层次的快乐建立在一条基本原则之上，即"在两种快乐之间，如果其中一种，所有或几乎所有经历了两者的人都对它做出偏爱的决定，不考虑偏好它的任何道德责任的感情，那么这种就是更令人向往的快乐"。[9]这也许看上去武断，因为它完全取决于人们的偏好，但是穆勒认为"聪明人不会同意当傻瓜；受过教育的人不会是无知的人；有感情和良心的人不会自私和卑鄙"。[10]因此，穆勒得出一个著名结论，"做一个不满足的人胜过做一头满足的猪；做不满足的苏格拉底胜过做一个满足的傻瓜。"[11]与"猪的哲学"相反，穆勒认为功利主义尊重人类生活的尊严和所有个体幸福的追求。

语言表述

穆勒是为普通读者撰写《功利主义》这本书的。就像英国政治哲学家阿兰·瑞安[*]所说，他的读者不是受过训练的逻辑学家。[12]穆勒以清晰的形式展示了对功利主义的常见反对意见，并依次分别对其作出回应。文本引人入胜，辩论的特征和观点清晰突出。他还对概念进行定义并提供支持功利主义的理由。

　　然而，解读《功利主义》一直是困难的。穆勒的同时代人，包括哲学家乔治·爱德华·摩尔*和佛朗西斯·布兰德雷*，认为该书前后矛盾。然而，现代学者如亨利·韦斯特和大卫·里昂，则另有一番见解。穆勒的观点对伦理和政治学领域的学生和学者而言很有价值，因为他所探究的概念，诸如功利主义、安全、正义、自由，到今天依然与我们息息相关。关于道德和自由主义本质的学术争论经常参考引用穆勒的观点，许多学术书籍致力于分析他的思想。政治家和政策制定者等在内的许多人也会基于功利主义的考量来进行决策，尽管很难分辨他们借鉴的是穆勒的具体观点还是一般意义上的功利主义思想。

1. 约翰·斯图亚特·穆勒，罗吉尔·克里斯普编：《功利主义》，牛津：牛津大学出版社，1998年，第2章。

2. 穆勒：《功利主义》，第55页。

3. 穆勒：《功利主义》，第55页。

4. 穆勒：《功利主义》，第55页。

5. 穆勒：《功利主义》，第57页。

6. 穆勒：《功利主义》，第63页。

7. 穆勒：《功利主义》，第63—64页。

8. 穆勒：《功利主义》，第64页。

9. 穆勒：《功利主义》，第56页。

10. 穆勒：《功利主义》，第56—57页。

11. 穆勒：《功利主义》，第57页。

12. 阿兰·瑞安：《约翰·斯图亚特·穆勒》，伦敦：劳特利奇&基根·保罗出版社，1974年。

6 思想支脉

要点

- 穆勒主张我们依据功利主义原则行事的动机是出于良知和对他人的同情。

- 穆勒宣称正义与作为道德哲学的功利主义完全相容，而安全对功利主义而言非常重要。

- 尽管有关正义与功利之间关系的章节在全书中所占篇幅最长，文献资料给予的关注却相对较少。

其他思想

约翰·斯图亚特·穆勒在《功利主义》一书中将幸福确认为人生的目标并定义了功利原则，他随后继续解决一系列更深层的问题。其中三个主题特别值得关注：动机、正义、安全。

第一个是人们按道德正确行事的动机。这个问题之所以重要是因为它解决了功利主义立场上的矛盾。一方面，穆勒说我们自己的幸福是生活的目的（即目标）；这是所有人都追求的。另一方面，他宣称道德上正确的行动经常需要个体为了共同利益而牺牲自己的幸福。如果第一个主张是正确的，那么我们第二个主张行动的动机从何而来？穆勒的回答是我们的良知和对人类的同情激励我们去这样做。或者，如他所述，我们是受到了"人类的社会情感、与我们的同类团结的愿望"[1]的激励。

第二个主题是功利主义可能与正义相冲突的担忧。功利主义在于最大化最多数人的幸福。但是批评者主张，为了实现这一点，有时一些人的利益可能会因为其他人而被牺牲。这是不公正的。举例来说，为了阻止愤怒的群体骚乱，可能不得不将一个无辜的人关入监狱。这样的监禁将产生更大的总体幸福（它阻止了骚乱），但是它对那位无辜的囚徒而言是不公正的。穆勒用他的最后一章证明正义对功利主义来说至关重要，而且与功利主义观点毫不冲突。

穆勒在《功利主义》中阐述的第三个观点有关安全。尽管《功利主义》主要关注幸福，穆勒认为安全感对人们感受到幸福来说极其关键。他将政治和法律的要素引入书中，主张我们应该为给予人们安全感创造更好条件。

> "因此，所有道德（除了外部动机外）的最终裁决是我们自己头脑里的主观感受，我看不出那些以功利为标准的人们有何尴尬可言，在这个问题上，什么是那个特定标准的裁判？我们可以回答，与其他所有道德标准一样：人类的良知。"
>
> ——约翰·斯图亚特·穆勒：《功利主义》

思想探究

穆勒主张良知和对他人的同情激励我们按照道德而行动。他相信道德并不取决于外在的约束，例如法律、社会的惩罚或上帝的非难。他认为所有道德的最终约束是"我们自己头脑里的主观感受"——即"人类的良知"。[2] 这从穆勒的角度来说是巧妙的推理，因为他现在可以宣称通过帮助他人，我们实际上服务于自己的幸福。他写道，"只要人们互相合作，他们的目的就是相一致的；至少会有一种暂时的感觉，其他人的利益就是自己的利益。"[3] 当为了他人的普遍幸福牺牲自我时，我们满足了良心的需求。这给予我们强烈的快乐，并解释了我们如此行事的动机。

穆勒认为功利主义与正义不对立，相反，正义可以用效用来解释。他总结道，正义感基于的信念是，人们的行为、社会机构、社会政策不应该损害我们自己或其他任何人的幸福。用穆勒的话说，"在我看来，正义感就是对自己或那些自己同情的对象受到的伤害或损害进行驱除或报复的动物欲望，通过人类放大同情的能力和智慧的利己观念，将这种情感扩散到包括所有人在内。"[4]穆勒认为正义与功利主义的原则不对立。两者互相兼容，而且功利主义依赖于正义，正义以功利主义为根基。

这使我们直接把握了第三个主题——穆勒认为安全如此重要的原因。穆勒将安全视作所有人都必需的东西，并称其为"所有利益中最重要的"。[5]尽管人们渴望各种事物，他们都对进一步保障安全感兴趣。既然安全是人人都享有的共同利益，那么为所有人最大化安全理所应当。因此，增进安全符合所有人的利益，因为从长期来看个人利益也可以得到照顾。当安全无法得到保障时，我们就会认为不公正，尽管道德安全的重要性到最后是通过功利——最终的善来解释的。

被忽视之处

约翰·斯图亚特·穆勒的《功利主义》是一部被一再仔细研读的简短著作。人们以为其中没有任何部分被忽视。然而，大多数分析者集中于第二章和第四章的特定段落，而未对整本书进行解读。最后一章对正义与功利主义的关系的考察，受到的关注远远少于试图证明功利原则正确性的第四章。

　　这可能归因于早期对《功利主义》的回应集中在穆勒对功利原则的证明上。思想家们，包括英国哲学家佛朗西斯·布兰德雷和乔治·爱德华·摩尔，试图证明穆勒犯下了若干逻辑谬误*。现代学者如亨利·韦斯特、大卫·里昂、乔纳森·里雷、约翰·斯科鲁普斯基，继续从这些针对穆勒功利原则证据的早期反对意见中汲取养分。因此，有关证据的一章一直以来最为学者所关注。

---

1. 约翰·斯图亚特·穆勒，罗吉尔·克里斯普编：《功利主义》，牛津：牛津大学出版社，1998 年，第 77 页。

2. 穆勒：《功利主义》，第 75 页。

3. 穆勒：《功利主义》，第 78 页。

4. 穆勒：《功利主义》，第 97 页。

5. 穆勒：《功利主义》，第 98 页。

7 历史成就

要点 ⚷

- 《功利主义》发展了社会改革家和哲学家杰里米·边沁的功利主义观点，澄清了功利主义是什么，并反驳了对功利主义的通常批评。
- 该著作将功利主义推向道德哲学里举足轻重的位置，并跻身于道德哲学最伟大著作的行列。
- 但是，该著作看似存在逻辑谬误：它没能澄清高级和低级的快乐如何会影响总体幸福的计算，也未明确穆勒支持"规则功利主义"*还是"行为功利主义"*（有关如何定义"正确"的行为的两种途径）。

观点评价

约翰·斯图亚特·穆勒在《功利主义》中的主要目标是阐明功利主义在道德哲学中的地位，发展思想家杰里米·边沁的功利主义观点，直面批评并为自己的功利主义进行辩护。在这方面，这本书收效斐然。穆勒比以往任何人都更清楚地陈述了何为功利主义，而他对高级快乐和低级快乐的区分对道德哲学的争辩影响深远。对于功利主义的批评者，他的回应详尽而富有洞察力。他在第四章对功利主义的证明尽管存有争议，却依然备受关注，使得功利主义成为当今道德哲学中最受支持的观点。

《功利主义》既是道德哲学最重要的书籍之一，也是最广为阅读的对功利主义观点的陈述。在高校，它被广泛用作向学生介绍道德哲学或伦理学的入门读物。这部著作的核心论点一直有各种解读，虽然在一些议题上迄今没有达成共识。学者们继续为妥当解读穆勒的论点和哲学立场而争辩不休。

> "《功利主义》是道德哲学最重要的著作之一，在重要性上与亚里士多德的《尼各马可伦理学》和伊曼努尔·康德的《道德形而上学原理》并列。"
> ——罗吉尔·克里斯普：《劳特利奇哲学指南之穆勒〈功利主义〉》

当时的成就

穆勒写作此书时，曾在道德哲学中备受推崇的功利主义日渐式微。事实上，边沁的功利主义面临严肃的批评，并被同时代的大多数知识分子质疑。穆勒的著作对巩固功利主义的立场至关重要，使得功利主义在哲学和日常政治生活里都成为一股重要力量。

穆勒攻击伦理直觉主义——这个概念粗略概括来说，就是有可能只通过直觉来理解伦理行为——而且，他说功利原则才是道德的基础，而非上帝的旨意。他也对边沁版本的功利主义进行了修正。因此，他会饱受许多不同思想家的批评，特别是直觉主义者、其他功利主义者以及宗教学者的批评，也就不足为奇了。

尽管存在这些批评，穆勒的著作仍被认为是现存对功利主义的最好的表述。[1]当爱尔兰历史学家和政治理论家威廉·莱基[*]没有在他的《欧洲道德史》（1869）里提及穆勒而引起争议时，即便是功利主义的反对者也对他颇有微词。他们觉得穆勒的功利主义是该信条的最好版本，因此值得关注。[2]据美国学者杰瑞米·斯尼温[*]所说，这起争议使"有一点变得大致清晰，即要谈功利主义，就必须谈及穆勒版本的功利主义"。[3]

局限性

穆勒《功利主义》的一个局限性是它似乎依赖于若干谬误 *。从幸福是最大的善的主张出发，得出因此我们应该增进总体的善被认为是一种错误的推断。他被指控犯了"合成谬误"，这种谬误表现为将部分的真实推断为整体的真实（例如，"自行车的各部分都很轻，因此自行车很轻"）。因此，尽管个人的幸福是我们自己最终的善或许是正确的，但并不能进而得出总体的幸福是所有人最终的善。

第二个局限在于，穆勒除了声明高级快乐应该得到更大的权重之外，没有说明高级快乐和低级快乐之间的差异应该怎样被计算。根据边沁版本的功利主义，每种快乐都有等同的价值，每种快乐的个体单位（有时称作"功利单位"）可以被计算并加在一起。但在穆勒的功利主义观点中计算究竟该如何操作还不清楚。

第三个局限是穆勒同时为现在已知的"规则功利主义"和"行动功利主义"辩护。根据行动功利主义，一个行为在道德上的善取决于那个行为对总体幸福量的影响程度。规则功利主义则相反，一个行为在道德上的善取决于它是否符合对总体幸福量有积极影响的普遍规则。这一区别很重要，因为它可导致判断一个行为正确与否的矛盾主张。例如，杀戮有时可以大大提高幸福的总和，但杀戮在一般意义上并非如此。穆勒并未明确他支持何种版本的功利主义。他在著作前部分给功利主义下定义的时候写道："行动若倾向于增进幸福，就是正确的；倾向于造成幸福的反面，就是错误的。"[4] 但几页纸之后，他写道："人类行动的目的（即目标），也必然是道德的标准；相应地可被定义为人类行为的规则和戒律，通过遵守这些规则，所有人都有最大的可能过上上述生活。"[5] 近年来，英国伦理学学者罗吉尔·克里斯普 * 倾向于以行动功利主义的路线来解读穆勒。[6] 其他人偏好被规则功利主义所突显的部分。[7] 还有一些人认为两方面都可以在穆勒的著作中找到。[8]

---

1. J. B. 斯尼温："关于穆勒《功利主义》的若干批评，1861—1876 年"，约翰·罗伯逊和米歇尔·莱恩编，《詹姆斯·穆勒与约翰·斯图亚特·穆勒：百年纪念会议论文集》，多伦多和布法罗：多伦多大学出版社，1976 年。

2. 斯尼温："关于穆勒《功利主义》的若干批评"，第 40 页。

3. 斯尼温："关于穆勒《功利主义》的若干批评"，第 40 页。

4. 约翰·斯图亚特·穆勒，罗吉尔·克里斯普编：《功利主义》，牛津：牛津大学出版社，1998 年，第 55 页。

5. 穆勒：《功利主义》，第 59 页。

6. 参见罗吉尔·克里斯普：《劳特利奇哲学指南之穆勒〈功利主义〉》，伦敦：劳特利奇出版社，1977 年，第 102 页。

7. 参见戴尔·米勒：《约翰·斯图亚特·穆勒：道德、社会与政治思想》，剑桥：剑桥大学出版社，2010 年。

8. 参见亨利·韦斯特：《穆勒伦理学导论》，剑桥：剑桥大学出版社，2004 年。

8 著作地位

要点 ⚷⸺

- 穆勒的道德哲学在《功利主义》一书中得到了最清晰和最系统性的表述。但是许多观点在穆勒此前的论文和出版物中已现锋芒。

- 穆勒的著作展现了他一生的思想非常一致，揭示了他对个人自由和增加所有人幸福的明确坚守。

- 《功利主义》是穆勒所有著作中最重要的一部，它在当下的意义与刚发表时相比分毫未减。

定位

约翰·斯图亚特·穆勒早年即开始写作，在 20 岁之前已为报纸和期刊撰写了不少高水平文章。在 1861 年《功利主义》出版时，他已经完成了若干论文和书籍，并被认为是英国杰出的哲学家之一。他出版的首部主要作品《逻辑学体系》（1843）偏向主张将逻辑作为举证方法，并把直觉主义者作为攻击目标。1848 年，穆勒出版了《政治经济学原理》，这是经济理论上的重要著作。

穆勒的论著《论自由》（1859）为个人自由辩护，主张对他人造成伤害的行为应该受到限制。他在《功利主义》的第五章回到这个主题。穆勒称社会为了阻止对他人造成伤害而干涉个人行为是完全适当的。在《论代议制政府》（1861）中，他赞成在英国建立一种既是代议制又有更多公众参与的政府形式。他说公众的投票"保证了舆论将被用来维护和引导公众利益"。[1] 在《论妇女的屈从地位》（1869）中，他赞成妇女的解放 * 和性别间的平等。这些论述展现了穆勒眼中包括两个部分的自由社会："社会的维持需要创造性的个体和意志的统一。"[2] 这两点也展现了根植于功利主义的自由主义信条的发展。如学者罗吉尔·克里斯普在他关于穆勒作品的概述中总结所说，"《功利主义》里所讨论的许多问题在穆勒早期的论文中都已有预示。"[3]

> "《功利主义》里所讨论的许多问题在穆勒早期的论文中都已有预示：伦理和伦理理解的基础，首要和次要原则的重要性，功利主义的证据，人类幸福的源泉，道德动机和道德的'约束'。"
>
> ——罗吉尔·克里斯普：《劳特利奇哲学指南之穆勒〈功利主义〉》

整合

尽管穆勒就一系列不同话题写作，他的著作展现了显著的统一性和一致性。他的《逻辑体系》写作时间早于《功利主义》近20年，穆勒在书中抨击了直觉主义者。他论证道如果一个人在伦理上是个直觉主义者，那么按照逻辑此人在科学上也应该是个直觉主义者，并且接受我们可以不借助观察而发现自然规律的说法。他认为这样的立场荒谬无理。他在《功利主义》中对直觉主义 * 的攻击是在人类所有知识领域推进经验主义的长期努力的一部分，在他此前的作品中有明显预示。

在《论自由》（1859）里，穆勒为个人自由辩护：他主张人们应该被允许按照他们认为合适的方式行动，除非他们意图伤害他人。这一例外就是所谓的伤害原则，表述为"违背本人意愿、正当地向一个文明社群的某位成员行使权力的唯一目的，是阻止对他人的伤害"。⁴有两个主题再次在《功利主义》中浮现，穆勒主张，第一，道德的正确性体现在最大化快乐和最小化痛苦及苦难。这意味着人们不应该受到阻碍，除非他们伤害到他人并由此减少幸福的总和。第二，穆勒坚信每个人的价值都是平等的，每个个体的幸福都很重要。一个群体不能权当其他人是自己获取利益的工具。幸福总和的最大化意味着每个人都对幸福有平等的诉求。

最后，在他的《论妇女的屈从地位》（1869）里，穆勒拓展了他的论证以表明妇女应当获得与男子平等的权利。需要再次强调的是，最大化所有幸福的总和是启迪穆勒的主要目标。在他看来，这应当把男女都包括在内，由此可见他对19世纪英国妇女解放的热切恳求。

意义

约翰·斯图亚特·穆勒两部最知名的著作是《论自由》和《功利主义》。《功利主义》是对穆勒道德哲学最系统和最精确的表达，汇集了若干发轫于他年轻时的主题。罗吉尔·克里斯普认为"若把穆勒视为福音传教士，1861年以三篇论文系列的形式首次出版的《功利主义》可以被看作他的圣经"。⁵

穆勒生前就已作为英国最重要的哲学家之一而闻名。然而，《功利主义》不是为少数知识分子精英而写。此书在1861年以三篇系列论文的形式刊载在《弗雷泽城市与乡村杂志》上，一份1830年创办于伦敦的通俗的文学类期刊，读者群定位为中产阶级。《功利主义》首次以书的形式出版是在1863年，直到这时穆勒才增加了关于功利与正义的章节。

穆勒的声誉在过去150年间几乎没有什么变化。他仍然被认为是一位哲学巨人。大多数政治学、经济学和道德哲学的大学课程广泛地探讨穆勒的观点，特别是《功利主义》中表达的观点。这些观点依然一针见血、切中时弊，其重要性在近期也不太可能减弱。

---

1. 罗伯特·德维尼：《改革自由主义：约翰·斯图亚特·穆勒对古代、宗教、自由和浪漫道德的使用》，康涅狄格州纽黑文：耶鲁大学出版社，2006 年，第 218 页。
2. 德维尼：《劳特利奇哲学之改革自由主义》，第 219 页。
3. 参见罗吉尔·克里斯普：《劳特利奇哲学指南之穆勒〈功利主义〉》，伦敦：劳特利奇出版社，1977 年，第 12 页。
4. 约翰·斯图亚特·穆勒：《论自由》，伦敦：企鹅出版社，2010 年，第 21 页。
5. 克里斯普：《劳特利奇哲学指南之穆勒》，第 7 页。

第三部分：学术影响

9 最初反响

要点 🔑

- 早期对穆勒《功利主义》的回应大多数是敌意的。他被指责作了谬误的——即不合逻辑的——推论，而且没有坚定地站在功利主义立场。

- 穆勒没有根据这些批评修改他的论点。相反，他尝试着澄清他的实际主张。

- 在穆勒生前人们对他的主张争论不休，没有达成共识。实际上，关于正确解读穆勒观点的辩论持续至今。

批评

在出版后的头 15 年，约翰·斯图亚特·穆勒《功利主义》得到的反响大多数是抱有敌意。批评者们集中于他的论证中的所谓的逻辑缺陷。[1] 有四点被特别提出——"他关于道德概念的可导性来自非道德的概念；他的功利原则的'证明'；他对高级快乐和低级快乐的区分；他的常识道德准则可作为功利主义伦理中间公理（即被接受的事实）的说法。"[2] 早期批评家如乔治·爱德华·摩尔质疑穆勒对功利原则的证明是否为谬误。[3] 穆勒试图通过以下论证来证明功利原则：

"唯一能够证明物体可见的证据是人们实际上看到它。唯一能够证明声音可以被听见的证据是人们听到它……同样的，我想能证明某个事物是被渴望的唯一证据，就是人们确实渴望它。"[4]

摩尔指控穆勒在"值得渴望"和"可见"之间作了错误的类

比，提出反对的理由，"这一步的谬误如此明显，他竟没能看出来相当令人惊讶。事实上，'被渴望'并不像'可见'意味着'有能力被看见'那样，意味着'有能力被渴望'。被渴望只是意味着什么应该被渴望或值得被渴望。"[5]

一位有影响力的哲学家托马斯·H. 格林*则反对快乐是唯一的渴望对象的说法。他指出，欲望的对象和由于满足欲望而得到的快乐是不相同的。[6] 例如，当我们饥饿时，我们追寻的不是快乐而是食物。因此在这个例子里，快乐不应被视为欲望的对象。摩尔、格林和哲学家佛朗西斯·布兰德雷是早期批评穆勒区分定量和定性快乐（即以量来衡量快乐和以质来衡量快乐）的人，并且他们质疑他对快乐主义*原则的运用。[7]

必须说明的是，并不是所有人都反对穆勒的功利主义论辩。例如，苏格兰哲学家詹姆士·塞思*就曾在 1908 年发表的一篇文章里坚决支持穆勒的立场。[8]

> "（当）我说总体幸福对所有人的集合而言是善，我指的并非是每个人的幸福对其他人来说是善，虽然我想在社会和教育良好的状态下，应该如此。"
>
> ——约翰·斯图亚特·穆勒：《约翰·斯图亚特·穆勒的信件》

回应

到 1871 年为止，《功利主义》历经了四个版本，有相对较小的修改和增加。哲学家亚历山大·贝恩*，在 1882 年题为《约翰·斯图亚特·穆勒》的著作里，说他"不知道重印（《功利主义》第一版）的书卷有何变化，尽管它在《弗雷泽》出版时已饱受严厉批评"。[9]

这暗示着穆勒在随后的版本里没有对关于他的著作的最初批评作出回应。然而，美国学者杰瑞米·斯尼温对这部著作初期的受关注程度提出异议。他宣称在起初的七、八年里，《功利主义》没有受到太多公开发表的关注，纵然对这部文本的哲学反馈大部分为负面。[10]

穆勒的论证"幸福是好的：每个人的幸福对其本身而言是好的，因此总体的幸福对所有人的集合而言是好的"，[11] 是引起很多争论的诱因，并导致合成谬误的指控。穆勒意识到了这个批评，并在一份标注日期为 1868 年 6 月 13 日的写给记者的信件里，明确了他的立场："我只想……论证既然甲的幸福是善，乙的是善，丙的是善等等，这些幸福的总和一定是善。"[12]

穆勒的信写在佛朗西斯·布兰德雷的批评论文发表在他的《伦理研究》上之前。但是布兰德雷没有对此作出回应，而是反驳了穆勒在《功利主义》中的论证。需要指出的是，穆勒在信件里仍然坚持这个普遍原则，即如果把好的事物放在一起，作为整体的结果也是好的。因此，布兰德雷的批评看上去依然在理。[13]

冲突与共识

穆勒死于 1873 年。去世之后，他的反对者发动了对其功利主义的大规模攻击。在随后的岁月里，大量的支持和反对这些异议的论著涌现。穆勒在临终前写的信件显示他没有改变他的观点，而是认为他的大多数批评者只是没有理解他。而批评者如布兰德雷，没有注意穆勒所增加的澄清。相反，他们继续指出他们所认为的穆勒《功利主义》出版版本中的错误。

关于正确解读穆勒核心思想的辩论持续至今，很大程度上是在重复对穆勒核心论证的早期反响。然而，一些现代学者，其中包括亨利·韦斯特和大卫·里昂，提供了对于穆勒功利主义的修正性解读——一种质疑正统观点的解读——指出他的批评者误读了他。但是也有可能人们在为穆勒同过去以及现在的批评声争论时，把错误的观点归之于他。尽管如此，关于穆勒为功利主义的辩护的冲突注定将持续一段时间。

1. J. B. 斯尼温："关于穆勒《功利主义》的若干批评，1861—1876 年"，约翰·罗伯逊和米歇尔·莱恩编，《詹姆斯·穆勒与约翰·斯图亚特·穆勒：百年纪念会议论文集》，多伦多和布法罗：多伦多大学出版社，1976 年。

2. 斯尼温："关于穆勒《功利主义》的若干批评"，第 41 页。

3. 乔治·爱德华·摩尔：《伦理学原理》，剑桥：剑桥大学出版社，1903 年。

4. 约翰·斯图亚特·穆勒，罗吉尔·克里斯普编：《功利主义》，牛津：牛津大学出版社，1998 年，第 81 页。

5. 摩尔：《伦理学原理》，第 67 页。

6. 托马斯·希尔·格林：《伦理学导论》，牛津：克拉伦登出版社，2003 年。

7. 参见摩尔：《伦理学原理》；格林：《伦理学导论》；佛朗西斯·赫伯特·布兰德雷：《伦理研究》，伦敦：牛津大学出版社，1962 年。

8. 詹姆士·塞思："穆勒'功利主义'中的所谓谬见"，《哲学评论》第 17 卷，1908 年第 5 期，第 469—488 页。

9. 引用自斯尼温："关于穆勒《功利主义》的若干批评"，第 38 页。

10. 斯尼温："关于穆勒《功利主义》的若干批评"。

11. 穆勒：《功利主义》，第 81 页。

12. 约翰·斯图亚特·穆勒：《约翰·斯图亚特·穆勒的信件》第 2 卷，休治·艾略特编并作序，附玛丽·泰勒撰写的关于穆勒的生平记述，纽约：朗文、格林出版社，第 116 页。

13. 布兰德雷：《伦理研究》，第 113 页。

10 后续争议

要点 ☞

- 20 世纪下半期，功利主义成为道德哲学中的主力军。它得到了进一步的发展——但也受到严重的批评。

- 近数十年来，许多功利主义的新流派形成，包括"双重功利主义"*"规则功利主义"和"偏好功利主义"*。

- 有关功利主义的学术研究现在以理论和应用的形式常见于各个学术领域。

应用与问题

在 1873 年约翰·斯图亚特·穆勒去世的时候，曾经主导英国哲学思想并塑造了《功利主义》的经验主义已经失去了它的地位，先是让给唯心主义*（这种哲学立场，非常粗略地说，认为"现实"主要是观念问题），后来让给了逻辑实证主义*（这种哲学立场认为要在认知上有意义，一个论断必须在经验上可得到验证或在逻辑上可由经验上能验证的论断推导而来）。穆勒的社会哲学没有受到太多关注，直至 20 世纪下半期。此时的学者们，比如伦理学和法哲学专家卡尔·威尔曼*，开始回应关于穆勒功利原则证明的传统批评。自此，评论家如亨利·韦斯特、大卫·里昂和乔纳森·里雷[1] 提供了对于穆勒作品的修正性解读，包括对如何证明功利原则的解释。

20 世纪后半期，以穆勒的作品为兴趣中心，功利主义再次成为道德哲学的一支主力军。哲学家们近几十年来拓展穆勒的功利主义版本，形成新版本，找到支持的新论点，并进行细致改进。

与此同时，对功利主义新的批评也出现了。其中有三种在文献里尤为流行。第一种是"快乐机器"的想法，由富有影响力的政治哲学家罗伯特·诺齐克 * 提出。诺齐克争辩称如果我们有可能被连接到一台机器上，这台机器让我们永久地处在一种快乐的状态里，不管快乐是多是少，大多数人都不会将这描述为美好生活。[2]第二种是英国思想家菲利帕·富特 * 著名的有轨电车难题：拉动拉杆让迎面而来的火车杀死一个人而不是另外的五个人，似乎在道德上不同于把一个人推到火车前去拯救五个人的生命。尽管结果是一样的（一死五活），这个思维实验试图展现在道德评价中，重要的不仅仅是行动的结果（如穆勒所说），还有动机。[3]第三种，英国道德哲学家伯纳德·威廉斯 * 也提供了思维实验来展示根据最大幸福原则行事可能实际上非常不道德。所有这些实验都质疑功利原则是否能作为所有道德的基础。[4]

> "荒唐之处是要求……一个人，在由其他人的计划部分决定的效用网络总额面前，他应该放弃自己的计划和决定，并认可功利主义计算所要求的决定。"
>
> ——斯马特、伯纳德·威廉斯：《功利主义：支持与反对》

思想流派

自 20 世纪 60 年代以来，功利主义的若干不同流派逐渐形成。这里我们聚焦于如下三种流派：规则功利主义、双重功利主义和偏好功利主义。

1953 年英国哲学家詹姆士·奥普·厄姆森 * 发表文章主张穆勒在功利原则的基础上证明了道德规则的合理性（例如"不得偷

盗""不得杀戮")。[5] "规则功利主义"流派——基本观点是一个行为道德上的善取决于它是否服从一般性的增加幸福总和的规则——自此出现。这个观点的产生是为了更容易地计算一个行为可能引起多少幸福或不幸。相应的,该观点也受到大量的批评。

为了克服规则功利主义的局限,英国哲学家理查德·麦尔文·黑尔*发展了被称为"双重功利主义"的观点。根据该流派,尽管我们应该追随基于功利原则的规则,但在特殊情况下,我们可以打破这些规则行动。在特殊案例里,我们将采取行动功利主义的立场,并评估(不考虑一般性的规则)一个行动是否增进了幸福。[6] 双重功利主义因此结合了规则和行动功利主义。

第三种重要的流派是"偏好功利主义",最初由诺贝尔奖获得者经济学家约翰·海萨尼*发展而来,但主要是与澳大利亚道德哲学家彼得·辛格*联系在一起。[7] 该流派主张我们的动力并不是最大化快乐和最小化痛苦,而主要是满足我们自己的偏好。像穆勒一样,偏好功利主义宣称好的行为促进好的结果。但与穆勒不同的是,他们相信当人们被允许满足自己的偏好时,结果才会是好的。

当代研究

功利主义在当前学术中的重要性再怎么夸大也不为过。功利主义在哲学、心理学、经济学和政治学领域都是一支重要力量。几乎每个处理规范*问题(关注什么是被认为做事的正常或正确方式的问题)的人,不管问题是实际性的还是理论性的,都需要对此表明立场,即功利主义是否提供了对于行动的正确路径的恰当指导。一些学者致力于准确理解思想家如杰里米·边沁和穆勒在他们的定义、主张、论证中所指的意思为何。其他人则指出功利主义在总体上或它的特定版本中错在何处。还有人进一步拓宽了功利主义的视野，并像穆勒一样，继续面对批评者们为它辩护。

功利主义在伦理学应用中扮演了重要角色。例如，用功利主义的观点解读有关动物福利的辩论，动物能体验快乐和痛苦，因此显然可以主张整体的幸福或不幸应该包括动物在内。还有生命伦理学的辩论，功利主义可以帮助决定，例如什么样的病人在医疗中应该获得公共资金。其他领域诸如贫困也被从功利主义的角度进行了探讨。因此，功利主义的发展——还有批评——不止停留在理论层面，还体现在多种多样的应用形式上。

1. 参见乔纳森·里雷："穆勒非凡的道德理论"，《政治、哲学与经济》第 9 卷，2010 年第 1 期，第 67—116 页。

2. 参见罗伯特·诺齐克：《无政府、国家与乌托邦》，纽约：基础图书出版公司，1974 年，第 42—45 页。

3. 参见菲利帕·富特："堕胎的问题与双重效应原则"，《美德与恶行》，牛津：巴塞尔·布莱克维尔出版社，1978 年。

4. 参见 J. J. C. 斯马特和伯纳德·威廉斯：《功利主义：支持与反对》，剑桥：剑桥大学出版社，1973 年。

5. J. O. 厄姆森："穆勒道德哲学解析"，《哲学季刊》，1953 年第 3 期，第 33—39 页。

6. 参见 R. M. 黑尔："亚里士多德社会的进程"，《新系列》，第 73 期（1972—1973 年），第 1—18 页。

7. 参见约翰·海萨尼："道德与理性行为理论"，阿马蒂亚·森和伯纳德·威廉斯编，《功利主义及其超越》，剑桥：剑桥大学出版社，1982 年，第 39—62 页；彼得·辛格：《实用伦理学》，剑桥：剑桥大学出版社，1979 年。

11 当代印迹

要点 ⌘—

- 《功利主义》在哲学史上被视为一部经典著作。但它仍然启迪着当代的功利主义者，他们用穆勒的洞见支持他们的观点。

- 《功利主义》是道德哲学上的三大主要传统之一——并且穆勒强有力地论证了为什么其他两种也可以被简化为功利主义。

- 伴随着那些"结果化"的行动变得越来越重要，关于功利主义的哲学辩论仍在继续。

地位

约翰·斯图亚特·穆勒的《功利主义》既是道德哲学中最重要的书籍之一，也是最为广泛阅读的有关功利主义的著述。在高校，它被广泛用作向学生介绍道德哲学或伦理学的入门书籍。它是哲学史上的一部经典之作。当前的哲学家们仍然参阅该书，或是支持他们自己的功利主义理念，或是批评或捍卫已在穆勒的观点中得到表达的功利主义的各方面内容。现代功利主义者如澳大利亚道德哲学家彼得·辛格，相信自己正在传承穆勒捍卫功利主义的事业，也在继续引用这部著作。

在关于正义本质的辩论中，《功利主义》也仍然是一个重要的源头，主要是由于第五章中穆勒展示了他关于功利与正义间关系的看法。约翰·罗尔斯*，20世纪著名正义理论家，在批评功利最大化是正义的基础这一观点时，直接引用了《功利主义》。[1] 罗尔斯认为，为了让其他人受益而牺牲某个个体的利益（根据功利主义，在一些情况下这么做于道德而言是正确的）是不公正的，除非这个利益被牺牲的个体得到某种形式的补偿。穆勒的著作绝不是思想史上的陈旧经典。尽管功利主义传统有所发展，但这部著作在当前的道德哲学领域继续占据中心位置。

> "当前哲学上的潮流与过去一百年或更多时间相比，更容易欣赏穆勒、反思他的作品并加以运用。"
>
> ——约翰·斯科鲁普斯基，《剑桥穆勒指南》前言

互动

功利主义通常被视作道德哲学的三大主要传统之一；另外两种为康德的道义论（一种以义务概念为中心的道德哲学）和亚里士多德的德性伦理学（一种以美德行为培养为中心的道德哲学）。在《功利主义》里，穆勒提出对这两种传统的反对论点。他抓住康德所说的我们不应根据一条不能被普遍化的原则行动，"他向人们说明的仅仅是，没有人会愿意承担由普遍接纳的规则带来的后果。"[2] 穆勒对德性伦理主义者更有好感，特别是关于他们对教育和良好品格发展的强调。但同样，他认为良好的品格如此重要的理由可通过功利原则得到最好的解释：良好的品格将导向更好的行为表现。它也将导向更多的总体幸福，后者是我们生活的主要目标。

在他对这些传统的批评里，穆勒实际上将德性伦理学和康德道义论转化为了功利主义的形式。这种对康德道义论和德性伦理学的抨击方式近年来得到复兴并被称为"结果化"。[3] "结果化"，如这些辩论的一位贡献者写道，"就是拿来一种推定的非结果主义的道

德理论，然后说明实际上它只是结果主义的另一种形式。"[4] 结果主义者赞同穆勒，产生好的结果是最重要的。尽管功利主义面临困难，近年来已经证实它是对道德哲学中不以结果为关注对象的流派（比如意图、性格特质）的重大挑战。

持续争议

当代关于功利主义的辩论有很多形式。首先，当代哲学家们继续认同古典功利主义者如杰里米·边沁和约翰·斯图亚特·穆勒实际捍卫的立场。除了亨利·韦斯特、大卫·里昂和约翰·斯科鲁普斯基的贡献之外，值得提及的是伦敦大学学院弗莱德·罗森的一本新书，题为《古典功利主义：从休谟到穆勒》（2003）。[5] 其次，哲学家们继续辩论功利主义的有效性。来自罗伯特·诺齐克、菲利帕·富特和伯纳德·威廉斯的批评激起了当今许多哲学家的回应。

第三，道德哲学家们争论结果主义者对康德道义论和德性伦理学的某些抨击。许多康德主义者和德性伦理主义者试图回答为什么他们的具体观点不能被"结果化"。道德哲学教授克莉丝汀·科斯佳 * 在最近的论文中主张康德和亚里士多德的立场都不能被结果化，因为普遍的善，若不利于某个人，就是无法成立的。[6]

虽然这些辩论是最近产生的，而且毫无疑问需要些时间才能达成共识，但值得指出的是，"结果化的行动"在穆勒的著作中有其根源。正因如此，以及许多其他原因，道德哲学教授斯科鲁普斯基才认为"当前哲学上的潮流与过去一百年或更多时间相比，更容易欣赏穆勒、反思他的作品并对之加以运用"。[7]

1. 约翰·罗尔斯：《正义论》，马萨诸塞州剑桥：哈佛大学贝拉纳普出版社，1971年。

2. 约翰·斯图亚特·穆勒，罗吉尔·克里斯普编：《功利主义》，牛津：牛津大学出版社，1998年，第52页。

3. 坎贝尔·布朗："后果化'这个'"，《伦理学》第121卷，2011年第4期，第749—771页。

4. 布朗："后果化'这个'"，第749页。

5. 弗莱德·罗森：《古典功利主义：从休谟到穆勒》，伦敦：劳特利奇出版社，2003年。

6. 克莉丝汀·科斯佳："论行善"，《哲学：皇家哲学研究所期刊》第89卷，2014年第3期，第405—429页。

7. 约翰·斯科鲁普斯基："导论"，约翰·斯科鲁普斯基编，《剑桥穆勒指南》，剑桥：剑桥大学出版社，1998年，第2页。

12 未来展望

要点 &----

- 《功利主义》将继续发挥重要性，不仅是作为功利主义的导论性文本，而且作为该理论未来发展的丰富源头。

- 当代哲学家继续以穆勒的中心论证为基础为他辩护，驳斥那些反对其论点的批评。

- 《功利主义》是道德哲学上的重要著作，因为它是首个对功利主义全面系统的解释和论证，并有证据的支持。

潜力

我们必须认同约翰·斯图亚特·穆勒的《功利主义》仍将继续在道德哲学、经济学和政治理论方面发挥重要的影响。作为对功利主义清晰系统的表述，它将继续作为道德哲学最重要和最流行的理论之一的奠基性文本而存在。功利主义传统在过去几十年发展迅速。现在功利主义的几个不同流派争论如何根据行动、规则或偏好等定义功利原则。尽管近年来穆勒观点的快乐主义（即以快乐为导向的）方面让位于著名澳大利亚道德哲学家彼得·辛格和其他人提出的偏好功利主义，这部著作对现代道德哲学家们依然有着巨大的吸引力。

近年来在关于政治自由主义的文献中穆勒得到的关注较少。自由主义思想家转而把影响很大的 20 世纪道德哲学家约翰·罗尔斯的《正义论》作为兴趣中心，该著作将正义定义为对自由社会运转至关重要的公平。约翰·罗尔斯的著作在 1971 年的出版见证了大学对基于权利的政治哲学不断增长的兴趣。那时候，功利主义占据政治哲学的主导，然而，罗尔斯的著作重新审视了被功利主义取代的政府和被统治者间的"社会契约"*思想。数百本书和数以千计的论文回应罗尔斯的著作——这是政治哲学领域基于权利的政治哲学大体上取代了功利主义主导地位的证据。

自20世纪50年代以来，对穆勒《功利主义》的学术研究集中于对这部著作的正确解读。关于穆勒作品的二级文献总量巨大并继续增长。这部著作的核心观点很有可能会进一步得到发展，因为许多哲学家现在回归到穆勒关于人类幸福、功利原则、正义与功利间的关系的思想。在可预见的将来，这些会继续成为积极从事这部文本研究的哲学家们的主要兴趣。

> "由于它简短、可读、论辩性强和富有说服力，它总是为了解道德哲学的复杂性和功利主义运动信条提供一条简易途径。"
>
> ——阿兰·瑞安，《功利主义和其他论文》导论

未来方向

我们必须假定，在与它的最激烈的批评者的直接对话中，功利主义将继续得到发展。这再现了穆勒在对他那个时代的批评者们的直接回应中发展自己学说的方式。尽管穆勒的《功利主义》将在何种程度上促进这些发展难以预测，但有一些当代哲学家可被提及，他们有很大可能会继续发展穆勒的著作。

例如，英国哲学家阿兰·瑞安仔细分析了该书第五章，并将其中的深刻见解与穆勒《论自由》中的若干章节及《逻辑体系》中的

一些段落结合起来。瑞安总结出穆勒发展的自由主义信条以功利主义为基础，由此重新确立了政治自由主义与穆勒古典自由主义间的密切联系。[1]像瑞安这样的学者主张，如果将穆勒的《功利主义》与他的其他著作联系起来考察，可以更好地理解这部作品。

另一方面，英国政治哲学家约翰·格雷*和哲学学者弗莱德·伯杰*考察了穆勒有关幸福的中心观念，并试图展示它对穆勒理解权利、自由、正义、自主和道德规则的重要性。[2]伦理学者约翰·斯科鲁普斯基是穆勒作品的一位长期崇拜者，目前正在评估约翰·罗尔斯提出的对穆勒自由观念的批评。像这样正在进行的研究可能为穆勒的自由主义回归哲学与政治前沿铺平道路。

这些例子显示，《功利主义》绝不是过时的思想史上的装饰物。它仍然是功利主义道德哲学家们灵感的源泉，是对非功利主义者的一个挑战。

小结

约翰·斯图亚特·穆勒的《功利主义》在道德哲学史上是一部关键作品。它包含了对下面问题的系统而全面的回答："我们应该如何度过人生以及我们应该如何共处？"

穆勒结合了他的父亲詹姆士·穆勒和导师杰里米·边沁的深刻见解，以及来自古代哲学家们的伦理学深刻见解。他主张幸福是人生终极的善，因此一个行为如果倾向于增进人类幸福，并减少世界上不幸的数量，那么它就是正确的。穆勒的功利主义学说的吸引力来自他在经验基础上证明功利原则的努力，包括区分高级快乐和低级快乐，以及发展基于效用的正义观。

尽管穆勒的作品在当时饱受批评，它已经并将继续对道德哲学产生巨大影响。功利主义的后续发展建立在穆勒的深刻见解之上并经常从他的著作中汲取灵感。学者们也继续考察《功利主义》里的中心论证，试图展示它所面临的许多批评都是基于误读。

如何度过美好的人生和如何在这世上举止良好，对于任何对这些问题感兴趣的人来说，《功利主义》仍然是值得研读的最重要的文本之一。

---

1. 阿兰·瑞安："约翰·斯图亚特·穆勒与生活的艺术"，约翰·格雷和 G. W. 史密斯编，《聚焦穆勒的〈论自由〉》，伦敦：劳特利奇出版社，1991 年；约翰·瑞斯：《约翰·斯图亚特·穆勒的〈论自由〉》，牛津：克拉伦登出版社，1985 年；弗莱德·伯杰：《幸福、公正与自由：约翰·斯图亚特·穆勒的道德与政治哲学》，伯克利：加利福尼亚大学出版社，1984 年。
2. 约翰·格雷："穆勒论自由：一种辩护"，伦敦：劳特利奇 & 基根·保罗出版社，1983 年；伯杰：《幸福、公正与自由》。

术语表

1. **先验**：可以独立于经验之外获得的知识，如数学知识。

2. **行为功利主义**：一种道德理论，认为行为的道德美好性取决于该行为对幸福总额的影响程度。

3. **计算**：建立在计算的基础上。

4. **道义论**：建立在义务和道德责任基础上的一种伦理学方案，主要归功于 18 世纪德国哲学家伊曼努尔·康德。

5. **解放**：从法律、社会或政治的束缚中获得自由。

6. **经验**：建立在经验基础上，通常是感官经验。

7. **经验主义**：一种哲学观点，主张所有的知识都建立在与理智或直觉相对的感觉经验的基础上。

8. **伦理直觉主义**：一种主张我们拥有先验的道德真理知识的信条，道德通过直觉被知悉，直觉与特定对象的直接意识相关联。

9. **伦理理论**：关于人类应该如何行动或行为的理性解释。

10. **谬误**：导致论证或主张无效的推理错误。

11. **合成谬误**：把某一部分的真实性推断为整个事物的真实性时所产生的错误推理，例如，自行车的所有部分都很轻，因此自行车很轻。

12. **法国大革命**：在 1789 年到 1799 年之间，法国经历了动荡和革命，见证了君主制的终结和 1793 年处决国王路易十六。革命前夕，教会领袖和统治阶层行使相当大的权力并享有特权生活，而普通民众则面对贫困和高额税收。

13. **最大幸福原则**：一种伦理原则，认为一个行为只要是增进了受其影响的最大多数人的最大幸福，那么这个行为就是正确的。

14. **快乐主义**：主张幸福或快乐是最终的善的信条。

15. **快乐主义的**：依赖于快乐的经验的。

16. **唯心主义**：一种哲学信条，主张现实由观点、思想、想法组成，而非有形物体。然而重要的是，这个术语没有一个单一的含义。在英国，唯心主义是从 19 世纪中期到 20 世纪早期占据主导地位的一种哲学运动。

17. **归纳伦理学派**：一种宣称道德知识基于观察和经验的哲学流派。

18. **工业革命**：18 到 19 世纪时期，英国走向以工业生产为基础的经济，经历了经济、社会和技术的根本性的变化。

19. **直觉主义**：一种所有的知识都基于直觉的信念——即知识的获得无需求诸推理、观察或经验。

20. **直觉伦理学派**：一种指向道德直觉或感觉的存在的思想流派。对该流派而言，普遍的道德原则先验性地被获悉，即它们是自明的。

21. **法学家**：法学学者或法学理论家，研究法律理论。

22. **康德道义论**：道德哲学的一种理论，以义务和规则为中心。康德版本的道义论认为，我们必须只为义务而行动，在此他指的是我们必须根据道德律来行动。

23. **自由主义**：捍卫个人自由不受国家无限控制的政治运动。

24. **逻辑谬误**：使得论证或主张无效的推理错误。

25. **逻辑实证主义**：一种知识理论。根据该理论，可以被经验验证，或由可被经验验证的表述出发、经逻辑推断而来的表述在认知上有意义。

26. **道德责任**：除开个人或环境因素，一个人必须做或有义务去做的事情，例如"不可杀人"。

27. **规范**：与理想的标准或模型有关的，或基于被认为是正常或正确的做事方式的。

28. **哲学激进主义者**：一群要求在英国进行社会改革的功利主义思想家。

29. **实证主义**：一种主张科学知识必须直接或间接以感官经验为基础，或经试验证实的信念。

30. **偏好功利主义**：一种道德理论，主张行为的善取决于它是否增进了人们满足自己需求和偏好的能力。

31. **功利原则**：主张行动或行为只要可以增进幸福或愉悦，就是正确的，如果导致不幸或痛苦，则是错误的。

32. **浪漫主义**：一种19世纪的哲学、艺术、文学运动，盛行于德国，重视人性中的非理性因素如想象力和情感。

33. **规则功利主义**：一种道德理论，主张一个行为的道德上的善取决于它是否符合通常会对幸福的总量产生积极影响的规则。

34. **社会契约**：一种观点，认为合法的国家权力必须来自被统治者的认同。

35. **白板**：一个拉丁语词汇，经常被译作"空白板岩"。在哲学上它被用来描述这一观点，人们出生时没有心理内容和先天知识，所有的知识来自经验和感知。"空白板岩"的哲学概念源自约翰·洛克。

36. **双重功利主义**：道德哲学的一种立场，主张我们应该追随基于效用原则的规则，但在特殊情况下，可以打破那些规则。这些例外情况是当遵循该规则时会导致幸福总量的减少。

37. **德性伦理学**：古希腊人伦理观的标志，由此优秀品格的美德培养被认为最为重要。

人名表

1. 阿里斯提波（前435—前356），创立了昔兰尼哲学学派的古希腊伦理学思想者，苏格拉底的学生，他教导生活的目标就是通过保持控制逆境和繁荣来寻求愉悦。

2. 亚里士多德（前384—前322），柏拉图的学生，古希腊哲学家，其思想塑造了西方哲学。他的著作涵盖多种主题，包括语言学、物理学、诗歌、音乐、生物学、政治学和伦理学。

3. 亚历山大·贝恩（1818—1903），苏格兰哲学家、教育家，在现代心理学的发展上扮演重要角色。他也是约翰·斯图亚特·穆勒的激进追随者。他的主要著作之一是《情感与意志》（1859）。

4. 切萨雷·贝卡里亚（1738—1794），意大利哲学家、政治家和法学家。他最为知名的著作是《犯罪与刑罚》（1674），该书批判了死刑与酷刑。

5. 杰里米·边沁（1748—1832），被认为是现代功利主义之父。他是社会改革家、哲学家、法学家，提倡产生"最多数人的最大幸福"的刑事司法制度和国家机构。

6. 弗莱德·伯杰（1937—1986），加州大学戴维斯分校哲学教授，研究方向为伦理学和法哲学。

7. 佛朗西斯·布兰德雷（1846—1924），英国哲学家，因《表象与真实：一篇形而上学的论文》（1893）而闻名。

8. 托马斯·卡莱尔（1795—1881），苏格兰历史学家和哲学家，同时代著名的文学家之一。

9. 萨缪尔·泰勒·柯勒律治（1772—1834），杰出的英国浪漫主义诗人。

10. 奥古斯特·孔德（1798—1857），法国哲学家和社会理论家，创造了"社会学"这一术语。他发展了实证主义的原则。

11. 罗吉尔·克里斯普（1961年生），牛津圣安妮学院道德哲学教授、导师，其著作主要关注伦理学。

12. 伊壁鸠鲁（前341—前270），古希腊哲学家，因倡导伊壁鸠鲁主义闻名。伊壁鸠鲁主义是一种提倡追求快乐，特别是精神快乐，并将其视作最高的善的哲学流派。

13. 菲利帕·富特（1920—2010），英国哲学家，以她在伦理学上的成就，特别是为重振古代德性伦理所作的努力而闻名。她最知名的著作是《美德和恶习及道德哲学中的其他论文》（1978）和《自然的善》（2001）。

14. 约翰·格雷（1948年生），英国政治哲学家，数部哲学畅销书的作者，因分析哲学和思想史方面的成果而有名。

15. 托马斯·希尔·格林（1836—1882），杰出而有影响力的英国哲学家和政治理论家。

16. 理查德·麦尔文·黑尔（1919—2002），英国牛津大学哲学家，研究方向为元伦理学，他最知名的著作是《道德的语言》（1952）、《自由与理性》（1963）、《道德思考》（1981），他在这些著作中为"普遍规定主义"的观点辩护。

17. 约翰·海萨尼（1920—2000），美籍匈牙利裔经济学家、诺贝尔奖获得者，以博弈论著称。

18. 克劳德·阿德里安·爱尔维修（1715—1771），法国哲学家、慈善家，倡导唯物主义人性论，该理论主张我们的行为由环境决定。

19. 大卫·休谟（1711—1776），苏格兰哲学家、经济学家和历史学家，以其经验主义（一种认为所有的知识都来自感觉的观点）和怀疑主义（认为真正的知识完全不可获得的哲学立场）而闻名。

20. 弗兰西斯·哈奇森（1694—1746），爱尔兰裔苏格兰哲学家、格拉斯哥大学道德哲学教授，以捍卫道德意识论和道德情感主义闻名。道德意识论主张人类有一种道德观念，通过它可以认可或否定人类的行为。

21. 伊曼努尔·康德（1724—1804），一位重要的18世纪德国哲学家，在现代哲学的发展上扮演了重要的角色。他以哲学上最有影响力的著作之一《纯粹理性批判》（1781）和《道德形而上学的基础》（1785）而闻名。

22. 克里丝汀·科斯佳（1952年生），哈佛大学哲学教授，研究方向为道德哲学，以其关于康德和亚里士多德的著作及对功利主义的批评而知名。

23. 威廉·莱基（1838—1903），爱尔兰历史学家和政治理论家，以《欧洲理性主义的兴起与影响史》（1865）和《从奥古斯都到查理曼的欧洲道德史》（1869）而知名。

24. 约翰·洛克（1632—1704），英国哲学家、医学研究者、学者，被誉为古典自由主义之父。

25. 大卫·里昂（1935年生），波士顿大学法律和哲学教授，研究方向为伦理学和法理学。

26. 哈丽特·泰勒·穆勒（1807—1858），英国哲学家，妇女权益活动家，为功利主义运动作出了显著贡献。

27. 詹姆士·穆勒（1773—1836），苏格兰哲学家、历史学家、经济学家，功利主义思想流派的重要代表人物。他在教育、政府、监狱、殖民地等多个主题领域均有著述。他的杰作之一是《政治经济学要义》（1821）。他是约翰·斯图亚特·穆勒的父亲。

28. 乔治·爱德华·摩尔（1873—1958），一位有影响的英国哲学家，以首次出版于1903年的《伦理学原理》闻名。

29. 罗伯特·诺齐克（1938—2002），20世纪晚期最有影响力的政治哲学家之一，以《无政府、国家和乌托邦》（1974）闻名。

30. 柏拉图（前427—前347），最著名、最有影响力的古希腊哲学家之一。他的作品《理想国》主要关注社会、公正、个人等议题，被认为是有史以来最伟大的哲学著作之一。

31. 约翰·罗尔斯（1921—2002），美国政治哲学家，20世纪政治哲学重要思想家之一，个人权利的坚定倡导者。他以其《正义论》（1971）闻名，这是 20 世纪一部里程碑式的著作。

32. 大卫·李嘉图（1772—1823），英国政治经济学家，其著作为劳动力市场和国际贸易作出显著贡献。

33. 乔纳森·里雷，美国杜兰大学哲学和政治经济学教授，研究方向为功利主义，特别是约翰·斯图亚特·穆勒的哲学。

34. 约翰·罗斯金（1819—1900），被认为是维多利亚时期最伟大的社会评论家和艺术批评家之一。

35. 阿兰·瑞安（1940 年生），英国政治哲学家，牛津大学政治理论荣誉教授，他关于约翰·斯图亚特·穆勒的著述颇多。

36. 杰瑞米·斯尼温（1930 年生），美国学者、哲学家，现任马里兰州约翰·霍普金斯大学荣誉教授。

37. 詹姆士·塞思（1860—1925），苏格兰哲学家，在爱丁堡大学担任道德哲学教授达 26 年。

38. 彼得·辛格（1946 年生），澳大利亚道德哲学家，与普林斯顿大学和墨尔本大学有联系，研究方向为应用伦理学，支持偏好功利主义，以《动物解放》（1975）和极富争议的支持杀婴而出名。

39. 约翰·斯科鲁普斯基（1946 年生），圣安德鲁大学道德哲学荣誉教授，研究方向为伦理学、认识论和道德哲学，以《理性的领域》（2010）闻名。

40. 苏格拉底（前 470—前 390），古希腊哲学家，西方哲学奠基者，他的思想主要通过他的学生柏拉图传达后世。

41. 詹姆士·奥普·厄姆森（1915—2012），哲学家、古典学者，在牛津大学度过大部分职业生涯，研究方向为伦理学、古代哲学和乔治·贝克莱的著作。

42. 卡尔·威尔曼，圣路易斯大学哲学荣誉教授，霍滕斯和托拜厄斯·卢因杰出大学人文科学教授，研究方向为伦理学和法哲学。

43. 亨利·韦斯特，明尼苏达州马卡莱斯特学院哲学荣誉教授，研究方向为功利主义，特别是约翰·斯图亚特·穆勒的哲学。

44. 威廉·休厄尔（1794—1866），英国哲学家、科学家、神学家，反对经验主义。

45. 伯纳德·威廉斯（1929—2003），英国剑桥大学道德哲学家，被认为是 20 世纪英国最伟大的哲学家之一。

46. 威廉·华兹华斯（1770—1850），杰出的英国浪漫主义诗人。

WAYS IN TO THE TEXT

- John Stuart Mill (1806—1873) was a British philosopher, political activist, and Member of Parliament. A defender of individual liberty throughout his writings and political career, he developed the moral philosophy of utilitarianism.

- According to his work *Utilitarianism* (1861), happiness, the greatest good in human life, is produced by morally good action.

- *Utilitarianism* is a key text in moral philosophy—the branch of philosophy that inquires into ethics. It clearly explains the utilitarian doctrine and defends it against objections.

Who Was John Stuart Mill?

John Stuart Mill, the author of *Utilitarianism* (1861), was a British philosopher, economist, and civil servant. Born in 1806, he was the son of James Mill,* a Scottish philosopher and historian, and his wife, Harriet Burrow. James Mill taught his son from an early age about ancient philosophy and the moral theory of utilitarianism. In this he was helped by the philosopher and jurist* (legal scholar) Jeremy Bentham,* generally considered to be the founder of utilitarianism. Both men hoped that the young John Stuart Mill would himself become a believer in utilitarianism and carry on the tradition. He did.

Mill, a political activist and important social reformer, wrote many essays and philosophical works over his lifetime. His first major published text was *A System of Logic* (1843), in which he argues in favor of logic as a method of proof. In 1848, he published *Principles of Political Economy*, a major text in nineteenth-century economic theory. His 1859 essay *On Liberty* offers a defense of individual liberty and argues that conduct that causes harm to others should be limited. In 1863 he published *Utilitarianism*, in which he explains and defends the moral theory of his tutor, Jeremy Bentham.

Active at a time of major social change in Britain, Mill was well known in his own lifetime. For three years he was a Member of Parliament for the Liberal Party, a British political party with policies founded on belief in the importance of individual liberty; during this period, he campaigned to give women the right to vote. Mill married the philosopher and campaigner Harriet Taylor* in 1851 after two decades of friendship. A great thinker in her own right, she helped Mill to develop his philosophical, economic, and political theories.

What Does *Utilitarianism* Say?

Utilitarianism is a book dealing with moral philosophy, particularly the question of how we should act and how we should live our lives. In it, Mill argues that happiness is the goal of human life, and that things can only be considered as "good" if they promote human happiness. Mill thinks this first argument can be empirically* proven (that is, proven by making deductions from observable evidence); human beings, he points out, do in fact all try to attain happiness. He defines "happiness" as the experience of pleasure and the absence of pain and suffering (a definition that might be understood as "hedonistic").* The goal of human life is, therefore, the experience of pleasure and the absence of pain and suffering; this is also the ultimate good.

Although he takes this idea from his tutor and predecessor Jeremy Bentham, Mill points out that the idea was also present in the ethical works of the Ancient Greeks, especially the philosopher Epicurus.* He builds on this long tradition by distinguishing between higher and lower pleasures, recognizing that not all pleasures are equally relevant in terms of measures of happiness; in assessing happiness, he argues, more importance has to be given to higher pleasures, such as those we take from poetry, music, and the development of insight.

From the belief that happiness is the ultimate goal of human life, Mill then derives the most important thesis of the book— namely, that an action is morally good if it increases happiness and reduces suffering, and morally bad if it does the opposite. Importantly, it is not one's own happiness that morally good action promotes, but the total happiness of all those affected by one's actions. Mill writes that "the foundation of morals, Utility, or the Greatest Happiness Principle* ... holds that actions are right in proportion as they tend to promote happiness, wrong as they tend to produce the reverse of happiness."[1] This, in short, is the principle of utility.*

In *Utilitarianism* Mill defends this moral theory, even if it implies that one's own happiness might have to be sacrificed, and addresses many criticisms that his contemporaries raised against the principle of utility. Of these, the objection that it conflicts with our sense of justice is especially notable. In the final part of the book Mill discusses the relationship between justice and utility, arguing that justice is also based upon the principle of utility, and that therefore no conflict exists between what is just and what is morally right.

Why Does *Utilitarianism* Matter?

Utilitarianism is a key text in the history of philosophy. Its systematic and detailed defense of utilitarianism led to the principle becoming one of the three most important moral theories in Western philosophy.

Mill placed great emphasis on good moral education. His text, aimed at a general audience, aims to teach us how to think about moral questions in order to make us all better and happier people. We all wonder how we should act in some given situation, and we often ask ourselves, "What is the right thing to do here?" Mill's utilitarianism gives us an answer. The right thing to do, he writes, is the thing that brings about the *most* happiness and the *least* unhappiness. This is a far more practical principle than most other moral theories offered, and his defense of this idea might well lead to many of his readers thinking differently about the right way to act.

Utilitarianism is more than just an educational text, however; a foundational text of the utilitarian tradition, it both clearly defines and defends utilitarian principles. In Mill's time, many believed that utilitarianism was inadequate as a moral philosophy. But Mill's strong, comprehensive, and very intelligent defense helped utilitarianism become one of the most important theories in moral philosophy in the twentieth century. Despite the fact that many different strands of utilitarianism have been developed since Mill's time, they all build on Mill's insights and their arguments are backed up by this text.

Utilitarianism is therefore not only a founding text of one of the most important theories in moral philosophy—it continues to be highly relevant in debates about how we ought to act, and how we should live our lives.

---

1. John Stuart Mill, *Utilitarianism*, ed. Roger Crisp (Oxford: Oxford University Press, 1998), 55.

SECTION 1
INFLUENCES

THE AUTHOR AND THE HISTORICAL CONTEXT

KEY POINTS

* Mill's *Utilitarianism* is a founding text in the history of moral philosophy (the philosophical consideration of ethics). It developed the English social reformer and philosopher Jeremy Bentham's* utilitarian views into a refined and comprehensive moral theory.

* Mill was brought up to become a strong promoter of utilitarianism. Jeremy Bentham himself played a large part in his education.

* Mill wrote at a time of social and political change and responded to it by supporting liberalism* (a social movement founded to promote and defend individual liberty).

Why Read This Text?

John Stuart Mill, the author of *Utilitarianism* (1861), is one of the most influential thinkers and liberal reformers of the nineteenth century. The work, among his most widely read philosophical works, offers a definition and philosophical defense of the ethical theory* of utilitarianism, modifying and expanding the work of previous utilitarian theorists.

An "ethical theory" is a reasoned account of how humans ought to behave or act; according to utilitarian principles, we should promote the good (where good means more overall happiness than unhappiness). Although it is a theory associated with John Stuart Mill and the British philosopher and jurist Jeremy

Bentham, its core ideas did not originate with them.

Mill's *Utilitarianism*, in which we find a fully worked-out version of utilitarianism, still serves as one of the most important texts in moral philosophy. In it, Mill distinguishes between higher pleasures—poetry and the development of insight, for example—and lower pleasures, arguing that the former are more important in the total sum of happiness. At the heart of the utilitarian belief system he places human sympathy rather than the "greatest happiness principle"* put forward by his teacher Bentham (the latter being an idea more calculative*—that is, determined through calculation—in nature). Mill also argues at great length why we should act according to utilitarian morality.

With these modifications and expansions, Mill tries to defend utilitarianism against a number of objections that continue to be raised in literature on the subject today. The continuing interest in Mill's *Utilitarianism* is therefore not only due to its founding role in moral philosophy; the text also provides powerful answers to objections that are still made against utilitarianism.

"I went through the whole process of preparing my Greek lessons in the same room and at the same table at which [my father] was writing ... I was forced to have recourse to him for the meaning of every word which I did not know. This incessant interruption, he, one of the most impatient of men, submitted to, and wrote under that interruption several volumes of his History and all else that he had to write during those years."

——John Stuart Mill, *Autobiography*

Author's Life

Mill was born in London in 1806 and died in 1873. His father, James Mill,* agreed with the teachings of the seventeenth-century English philosopher John Locke,* who claimed that the human mind is a *tabula rasa**—a blank slate—at birth and that knowledge is acquired through experience. Wanting his son to continue the utilitarian political and social reform program after he and his friend Jeremy Bentham had died, James Mill embarked on a rigorous program of education in which he personally tutored his son in order to produce the perfect utilitarian mind.[1] In his autobiography, John Stuart Mill records how his father told him to study Greek and Latin, and made him systematically go through all the classical texts.

Mill records in his autobiography that he was required to study Ancient Greek "in the same room and at the same table" at which his father worked; "I was forced to have recourse to him," he wrote, "for the meaning of every word which I did not know."[2]

But this education did not immediately have the desired effect. In 1826, Mill suffered a severe mental crisis; his eventual recovery came through his engagement with poetry, and in particular with the work of the English poets William Wordsworth* and Samuel Taylor Coleridge,* which was typical of the Romantic* tradition in its emotionally oriented and descriptive qualities. Mill hoped to develop his aesthetic sensibilities by reading these poems.[3]

Following his recovery, he started to shape his own philosophical views. He emphasized the importance of emotions and feelings in human life and, in doing so, distanced his thought from the more calculative aspects of Bentham's utilitarianism.[4] Nevertheless, Mill remained committed to utilitarianism and empiricism* (the belief that knowledge should be based on experience) until the end of his life. In his autobiography, Mill credits his wife, Harriet Taylor Mill,* with helping him develop his philosophical thought.

Mill was also a political activist and served in parliament from 1865 to 1868 as a member of the Liberal Party. He was the first British parliamentarian to support the right of women to vote, and was a passionate and lifelong supporter of that cause.

Author's Background

Mill lived in a period when Britain was experiencing substantial social and economic changes as a result of the country's move from an economy based on agriculture and manual labor to one based on industrial production. This period, called the Industrial Revolution,* led to economic growth in Britain, but also resulted in a number of challenges, including an increase in inequality, rapid urbanization, unemployment, crime, rising prices, a massive rise in poverty, declining wages, poor housing, and a surge in illnesses. The nineteenth century was dominated by campaigns and debates on political reform (in part, inspired by the French Revolution* of 1789 to 1799, in the course of which the French king was executed and the government was radically reformed). There were loud demands for individual freedom, for democratic and parliamentary reforms, for a minimal state, for a free-enterprise economy, and for the need to establish a more tolerant and fairer society. During this period, there were also calls for the replacement of aristocratic rule with a representative government. In nineteenth-century Britain, the liberal political movement was opposed to aristocratic rule, its members believing that Parliament comprised a corrupt elite who acted in their own interests and violated individual liberty.

In short, Mill lived during a period of social upheaval. These challenges inspired much of his work and philosophical activity. He reacted to the changing conditions and advocated liberalism.

---

1. John Stuart Mill, *Autobiography*, ed. Jack Stillinger (London: Oxford University Press, 1971).

2. Mill, *Autobiography*, 9.

3. See Mill, *Autobiography*; Thomas Woods, *Poetry and Philosophy: A Study in the Thought of John Stuart Mill* (London: Hutchinson & Co., 1961).

4. See John Stuart Mill, "Bentham," in *Utilitarianism and Other Essays*, ed. Alan Ryan (London: Penguin Books, 2004).

MODULE 2
ACADEMIC CONTEXT

KEY POINTS

- *Utilitarianism* is a work in moral philosophy, offering an answer to the question how one ought to live.

- In it, Mill offers an alternative theory to classical virtue ethics* (an approach to understanding "right" action, dating to Ancient Greece, that focuses on the individual's cultivation of virtuous behavior) and deontology* (an approach to ethics founded on notions of duty and moral obligation* that owes much to the eighteenth-century German philosopher Immanuel Kant*).

- Although Mill was heavily influenced by his father and the Utilitarian philosopher Jeremy Bentham,* he also drew from the insights of other philosophers, both ancient and modern, and from contemporary critics of utilitarianism.

The Work in its Context

John Stuart Mill's *Utilitarianism* is a work in the field of moral philosophy, the area of philosophy concerned with the question "How should one live?"

This question, notably raised by the Ancient Greek philosopher Socrates,* can be split into three further questions: "What is happiness?", "How can we know what is morally right?", and "How are we motivated to do the morally right thing?"

Moral philosophy is concerned not with how things *are* in the world, but with how things *should be*. Rather than describing how people make their decisions and go about living their lives or analyzing how societies work, moral philosophy questions how people *should be* living their life, and how society *should be* organized. There has long been skepticism about the possibility of finding answers to questions in moral philosophy. Many believe that it is up to an individual to decide how they live their life. Most moral philosophers, however, argue that there is such a thing as a right action and a wrong action, a morally better or worse person, and more or less just ways of social organization. Mill's *Utilitarianism* presents an argument for one such view.

> *"In the present work I shall, without further discussion of the other theories, try to contribute something towards the understanding and appreciation of the Utilitarian or Happiness theory, and towards such proof as it can be given."*
>
> ——John Stuart Mill, *Utilitarianism*

Overview of the Field

Generally, there are three traditions in moral philosophy: virtue ethics, Kantian deontology,* and utilitarianism.

Virtue ethics is usually associated with the work of the Greek philosopher Aristotle,* the student of the enormously influential thinker Plato.*[1] In Aristotle's view, what determines the nature of human beings is their capacity to reason. To live a good life we must exercise our capacity to reason *well* (that is, to reason in accordance with "excellence"). In Aristotle's view, it is the "virtuous agent" (in other words, a person making excellent use of his or her capacity to reason) who will know how to act in specific circumstances. Rather than being a general principle determining

84

what is right and wrong, the virtue ethical position is that well-educated and virtuous agents will reliably choose the best course of action.

Kantian deontology, developed by the greatly influential German philosopher Immanuel Kant, by contrast, looks not at the character traits of the individual, but at the rightness and the wrongness of an action itself.[2] Kant tried to formulate a universal principle on the basis of "pure reason," paying no attention to what makes people happy. Kant's view is that morally good action consists in acting in accordance with a principle that one should be able to rationally *will* (that is, rationally want to come about) as a universal law.

So if one can rationally will a principle of action as a universal law, the action is morally good; if one cannot rationally will it (that is, if it were to result in contradictions), the action is morally wrong. This moral test is not about wanting to live in a world where everyone acts on certain principles. It is not the consequences that matter for Kant. What matters is that it must rationally be conceivable that the principle can become a universal law; if it ends up in a contradiction, any action based on the principle is morally wrong.

According to *Utilitarianism*, however, the goodness of an action does not depend on the character of the individual (as in virtue ethics); nor does it depend on the possibility of rationally willing it as a universal law (as in Kantian deontology). Instead, utilitarianism is the view that the goodness of an action depends on its *consequences*. If an action has good consequences, the action is morally good, whether or not it is performed by a virtuous individual. Utilitarians are only interested in the amount of overall happiness and unhappiness an action brings about. The morally right way to act, according to utilitarians, is the way that brings about the most good: it must increase the total amount of happiness in the world and decrease the total amount of unhappiness.

Academic Influences

The most direct influences on Mill's thinking were his father, James Mill,* and his tutor, the "founder" of modern utilitarianism, Jeremy Bentham. They were both part of the intellectual tradition known as "philosophical radicals,*" which also included writers such as the British economic theorist David Ricardo.* But the intellectual world of Mill's youth "combined openness to radicalism in the form of Jeremy Bentham's ... with a suspicion bred by his contact with ancients that not every idea of the good is reducible to gross pleasures."[3]

Thus, Mill's thought was also influenced by Ancient Greek philosophy. The idea that happiness is the ultimate purpose of human life, for example, was also present in the writings of ancient philosophers such as Aristotle,* Epicurus,* Aristippus,* and Plato.* The doctrine of utility had long been accepted by thinkers both before and during Mill's time; thinkers such as Claude Adrien Helvétius,* Cesare Beccaria,* John Locke,* David Hume,* and Francis Hutcheson* had all drawn on it to form ideas.

Opponents of utilitarianism included thinkers such as the Scottish philosopher, mathematician, and social commentator

Thomas Carlyle* and the English thinker John Ruskin.* Such critics did not view utilitarianism as the "fundamental principle of morality, and the source of moral obligation."*4 They believed it conflicted with what is actually morally right. Carlyle even claimed that utilitarianism, with its sole focus on pleasure, was in fact a "pig philosophy." Mill's arguments were influenced by these critical voices.

In *Utilitarianism,* Mill brings together the ideas of his father James Mill, his tutor Bentham, the insights offered by the ancient schools of philosophy, and arguments made by critics of utilitarian ideas.

---

1. See Aristotle, *Nicomachean Ethics*, in *The Complete Works of Aristotle, The Revised Oxford Translation*, ed. Jonathan Barnes (Princeton, NJ: Princeton University Press, 1995).

2. For Kant's moral philosophy, see Immanuel Kant, *Groundwork of the Metaphysics of Morals,* ed. and trans. Mary J. Gregor (Cambridge: Cambridge University Press, [1785] 1998).

3. Robert Devigne, *Reforming Liberalism: J. S. Mill's use of Ancient, Religious, Liberal, and Romantic Moralities* (New Haven, CT: Yale University Press, 2006), 2.

4. John Stuart Mill, *Utilitarianism*, ed. Roger Crisp (Oxford: Oxford University Press, 1998), 51.

THE PROBLEM

KEY POINTS

- Mill's *Utilitarianism* attempts to answer the fundamental moral question of how to live and act by defending and expanding the philosopher Jeremy Bentham's* philosophy of utilitarianism.

- Mill engages with all schools of moral philosophy in Utilitarianism—especially with those authors who raised objections to Bentham's views.

- Mill examined what utilitarianism stood for, how it could be defended against its critics, and why it could be supported by empirical* evidence better than other schools of ethics founded more on intuition.

Core Question

John Stuart Mill's *Utilitarianism* is primarily aimed at answering the question, "What is the right way to act or to live?"

In order to answer this question, Mill addresses a series of further questions central to moral philosophy. He asks, first of all, what the "end" or "goal" is of good action. Whenever we think of something as good, we do so because it results in something that we also think of as good. For instance: receiving a high grade at school is good because it might lead to graduate school, which is good because it helps the development of one's intellect, which is good for some further reason—and so on.

At some point this chain has to stop; there must be a *final* good (that is, something which means that everything that has contributed to it is good). Without a final good, we would not know what makes all the other things good. In order to know how to act, Mill thinks we should ask the question, "What does this final good consist of?"

Mill's answer to this question is happiness. But this in turn creates further questions. What does happiness consist of? What is the relationship between happiness and morality? What are we morally obliged to do if the final good is happiness? If happiness is the final good, what is the relationship between happiness and justice? These are the questions that concern Mill in *Utilitarianism*, and which he attempts to answer.

Mill contributes to the philosophical debate on how to act and live. It is the same debate that philosophers such as the Ancient Greek thinkers Plato,* Aristotle,* and Epicurus* and the eighteenth-century German philosopher Immanuel Kant* grappled with. The core questions of Mill's work are no different from earlier philosophical inquiry.

> "From the dawn of philosophy, the question concerning the [highest good], or, what is the same thing, concerning the foundation of morality, has been accounted the main problem in speculative thought, has occupied the most gifted intellects, and divided them into sects and schools, carrying on a vigorous warfare against one another. And after more than two thousand years the same discussions continue."
>
> ——John Stuart Mill, *Utilitarianism*

The Participants

In chapter 1 of *Utilitarianism*, Mill states that the followers of the "inductive school of ethics"* and the "intuitive school of ethics"* were in agreement that morality was based on the establishment of principles. The main participants Mill had in mind while writing the text, therefore, were defenders of these two schools. Although both schools agreed that morality required principles, they disagreed about the ultimate source of moral obligation* (that is, action that one has a moral duty to perform). In Jeremy Bentham's version of utilitarianism, pleasure and pain are the only things that govern human behavior, and the principle of utility* (or the greatest happiness principle)* is the basis of morality.[1] Followers of the intuitive school disagreed with this.

Mill also tries to combine the ideas of his father, James Mill, and Jeremy Bentham with those of other thinkers, including Thomas Carlyle* (a Scottish philosopher associated with the literary, artistic, and philosophical movement of Romanticism),* ancient philosophers such as Socrates* and Epicurus, and those such as the French social theorist Auguste Comte,* who subscribed to the tenets of positivism* with their belief that scientific knowledge must be derived from sensory experience and tested by means of experiments. The influence of Comte's positivism is clearly visible in Mill's work, as Mill sees the natural sciences as the only way to gain knowledge. Mill rejects the view held by thinkers such as the German philosopher Immanuel Kant and the British theologian and philosopher William Whewell* that *a priori** ethical principles—that is, ethical principles that can be deduced independently of experience—exist.

The Contemporary Debate

Although the contemporary debate was influenced by supporters of the inductive and intuitive schools of ethics, Mill thought there was also a great deal of misunderstanding about utilitarianism. As well as debates about whether utilitarianism was the *right* moral theory, there was also much discussion about what utilitarianism stood for in the first place. Mill addresses those who think that the promotion of utility in society means *not* promoting happiness. He points out that this is precisely what utilitarianism is about.

A second debate in Mill's time was whether utilitarianism was a kind of "pig philosophy," as Thomas Carlyle described it. Carlyle's critique was specifically aimed at Bentham's utilitarianism. Bentham believed that utility (or happiness) could be measured by adding up all the pleasures that people would experience. Critics argued that this form of utilitarianism was an animalistic moral theory, as it promoted the kind of gross pleasures that govern the actions of animals. Mill defends utilitarianism by arguing (against Bentham) that the *quality* of pleasures is of crucial importance; some pleasures are of a higher quality—and therefore of higher value—than others. This is contrary to Bentham's argument that pleasures that are equivalent in terms of intensity, duration, and certainty have equal value.

Mill also criticizes ethical intuitionism*—the doctrine that we have *a priori* knowledge of moral truths—as he sees it as just reflecting the status quo. He defends utilitarianism on an empirical* basis, saying that experience shows us people do in fact try to act in ways that promote their happiness, especially the kind of happiness that results from the "higher" pleasures.

---

1. Jeremy Bentham, *An Introduction to the Principles of Morals and Legislation*, eds. J. H. Burns and H. L. A. Hurt (London: Athlone, 1970).

MODULE 4
THE AUTHOR'S CONTRIBUTION

KEY POINTS

* In *Utilitarianism*, Mill tries to define the philosophy of utilitarianism and defend it against a number of criticisms, including the criticism that utility means something other than pleasure and happiness.

* *Utilitarianism* takes an empirical* approach to moral philosophy: in it, Mill claims that we can deduce that happiness is the greatest good by observing that we all clearly desire happiness for ourselves.

* Mill defends the greatest happiness principle* put forward by the philosopher Jeremy Bentham, but adds quality of pleasures to the measures of happiness.

Author's Aims

John Stuart Mill points out in the opening passages of *Utilitarianism* that critics of utilitarianism had grossly misunderstood its central doctrine. He goes on to list the many criticisms that had been leveled against utilitarianism: that happiness is actually unattainable; that everyone would end up dissatisfied if they acted on the "greatest happiness principle"; that utilitarian theory relies solely on cold calculation; that total happiness is too complex to calculate; that it is bound to result in conflicting duties; that it leaves no place for God in morality; that it just acts as a kind of self-justification of one's actions; that utilitarianism is a moral theory more suitable to animals than to human beings; and finally, that it conflicts with common ideas about justice.

But the first criticism Mill addresses in chapter 2 of *Utilitarianism* is the criticism that utilitarianism focuses on utility rather than pleasure. Mill responds to this misunderstanding by saying:"Those who know anything about the matter are aware that every writer, from Epicurus to Bentham, who maintained the theory of 'utility', meant by it, not something to be contradistinguished from pleasure, but pleasure itself, together with exemption from pain."[1]

Having made clear the meaning of utilitarianism, Mill tries to provide an empirical defense of it. He rejects other ethical positions such as ethical intuitionism,* and argues that morality is based on experience and observation. Mill is different from other utilitarians: he thinks that utilitarianism can be defended on *empirical* grounds (that is, through deduction based on observable evidence). The amendments he makes to the utilitarian doctrine itself also make him very different from earlier utilitarian thinkers.

> "Those who know anything about the matter are aware that every writer, from Epicurus* to Bentham,* who maintained the theory of 'utility', meant by it, not something to be contradistinguished from pleasure, but pleasure itself, together with exemption from pain; and instead of opposing the useful to the agreeable or the ornamental, have always declared that 'useful' means these among other things."
>
> ——John Stuart Mill, *Utilitarianism*

Approach

In chapter 1 of *Utilitarianism*, Mill first dismisses alternative theories, especially those that depend on ethical intuitionism; in chapter 2, he defines utilitarianism and considers the objections to it listed above; in chapter 3 he considers possible motives for people to follow the principles of utilitarianism; in chapter 4 he offers his proof of utilitarianism; and in chapter 5 he focuses on justice and its relation to utility. Mill does not simply defend utilitarianism against its opponents, but also refines and develops the way it was presented by James Mill* (his father) and Jeremy Bentham. He ends up with his own version of utilitarianism.

The originality of *Utilitarianism* lies in its central claims and its empirical justification. The empirical justification aligns his work with the positivism* of the French social theorist August Comte;* according to positivist thought, scientific knowledge must be based on sensory experience and tested by experiment. Mill argues that the empirical proof for utilitarianism is that people all desire their own happiness:

"No reason can be given why the general happiness is desirable, except that each person, so far as he believes it to be attainable, desires his own happiness. This, however, being a fact, we have not only all the proof which the case admits of, but all which it is possible to require."[2]

So as well as creating his own version of utilitarianism and defending it against objections, Mill provides an empirical proof of the doctrine. This was original and unorthodox. It also contradicted those philosophers who believed that moral intuitions were the source of morality.

Contribution in Context

Mill's tutor Jeremy Bentham, in his *Principles of Morals and Legislation* (1789), claims that human beings are created to seek pleasure and to avoid pain, and that they desire pleasure for its own sake. He says there are physical, political, religious, and moral sources of sanctions for pleasure and pain. The political, religious, and moral sources, however, are secondary to the physical sources. The greatest happiness can only be achieved by maximizing pleasure and minimizing pain. An action or law is therefore good if it produces the greatest amount of pleasure and the least amount of pain to everyone involved. Bentham developed a scientific and rational method that could be used to measure quantities of pleasure and pain. He then applied it to the English legal system and institutions.

Although Mill does not abandon Bentham's fundamental principle (that of utility), he criticizes his predecessors, including his father, for offering an inadequate notion of the good (i.e. more overall happiness than unhappiness). As such, Mill revises utilitarianism and expands the notion of utility to include *qualitative* pleasure. This is a departure from the utilitarianism of his predecessors.

This departure has led to claims that Mill is an inconsistent utilitarian. His aims and intentions in *Utilitarianism* are, however, clearly presented and form a coherent, well-argued plan, even if there have been many interpretations of his arguments defending the principle of utility.* He is charged, for instance, with making a logical error in his central argument when he claims that from the thesis that "each person's happiness is a good to that person" it follows that the general happiness is "a good to the aggregate of all persons."[3] Critics of this argument have charged him with committing a fallacy of composition* (that is, an argument founded on the incorrect deduction that what is true for a part is true for the whole). Modern writers on Mill's *Utilitarianism*, such as the philosophy scholars Henry R. West,* David Lyons,* Jonathan Riley,* and John Skorupski,* have offered alternative interpretations of Mill's argument in an attempt to unearth what he actually intended.

---

1. John Stuart Mill, *Utilitarianism*, ed. Roger Crisp (Oxford: Oxford University Press, 1998), 54.

2. Mill, *Utilitarianism*, 81.

3. Mill, *Utilitarianism*, 81.

SECTION 2
IDEAS

MAIN IDEAS

KEY POINTS

* Mill argues that the foundation of morality is that an action is morally right if it promotes overall happiness, and morally wrong if it promotes the reverse.

* To assess whether an action promotes overall happiness, Mill says, we should not only look at the quantity of pleasure it will bring about—we should consider the quality of the pleasures that will be promoted, too.

* Although *Utilitarianism* is written clearly and accessibly, there is still debate about the precise meaning of certain passages.

Key Themes

The key themes of John Stuart Mill's *Utilitarianism* are the principles of utility, higher and lower pleasures, and the development of good moral character. The central thesis of the book is stated clearly as "the doctrine which accepts as the foundation of morals, Utility, or the Greatest Happiness Principle, holds that actions are right in proportion as they tend to promote happiness, wrong as they tend to produce the reverse of happiness."[1] In defending this claim, Mill endorses his tutor Jeremy Bentham's* greatest happiness principle* as the foundation of morality.[2]

The ultimate goal of human life is happiness. Everything that people desire is desired because it will either lead to happiness or provide some happiness. Happiness must therefore be the ultimate good. Mill defines happiness as "intended pleasure, and the absence of pain." He defines unhappiness as "pain, and the privation of pleasure."[3] Mill argues that it is not only one's actions that should be directed toward maximizing happiness: one's character should also aim to achieve this goal. Although Mill does not agree with Aristotle's* belief that it is one's character that makes an action right, he does nevertheless think that having a virtuous character is important, as it will make one likely to act in ways that will promote happiness in the world.

Mill then goes beyond Bentham by making a distinction between the *quantity* of pleasures that will be experienced in society and their *quality*. This is a response to critics of utilitarianism such as the Scottish thinker Thomas Carlyle,* who wrote that to "suppose that life has ... no higher end than pleasure and no better and nobler object of desire and pursuit, they designate as utterly mean and groveling; as a doctrine worthy only of a swine."[4] Mill develops the crucially important idea that in addition to quantity, the *quality* of pleasures is important when measuring and ranking pleasures. Mill identifies two types of pleasure: higher pleasures (mental pleasures) such as poetry and knowledge; and lower pleasures (animalistic pleasures), such as sex and food. To find out which pleasure is qualitatively superior to another, one needs to check with those who have experienced both. Mill believes that individuals who have experienced higher pleasures prefer them to lower pleasures, and would not be willing to give them up even for "a promise of the fullest allowance of a beast's pleasures."[5] For Mill, higher pleasures are superior in quality and more desirable than lower pleasures.

"The doctrine which accepts as the foundation of morals, Utility, or the Greatest Happiness Principle, holds that actions are right in proportion as they tend to promote happiness, wrong as they tend to produce the reverse of happiness. By happiness is intended pleasure, and the absence of pain; by unhappiness, pain, and the privation of pleasure."

——John Stuart Mill, *Utilitarianism*

Exploring the Ideas

Having defined happiness as the end (that is, the goal) of human life, Mill does not conclude that we should simply try to increase *our own* happiness. Instead, he thinks we ought to promote *overall* happiness—that is, the sum of happiness experienced throughout a society. This implies that in an imperfect world where not everyone is happy, we sometimes ought to sacrifice our own happiness for the happiness of others. Mill considers the "readiness to make such a sacrifice is the highest virtue which can be found in man."[6] But unlike other moral theories, Mill adds that self-sacrifice *as such* is not good or heroic:"a sacrifice which does not increase, or tend to increase, the sum total of happiness, it [utilitarianism] considers as wasted."[7] Utilitarianism says that morally right action must promote the *general* good, not just one's own. Mill therefore compares it to the Christian doctrine "to do as one would be done by, and to love one's neighbor as oneself." This, in Mill's view, "constitutes the ideal perfection of utilitarianism."[8]

The higher and lower pleasures that Mill distinguishes in his version of utilitarianism are based on the basic principle that "of two pleasures, if there be one to which all or almost all who have experience of both give a decided preference, irrespective of any feeling of moral obligation* to prefer it, that is the more desirable pleasure."[9] This may seem arbitrary, as it relies completely on what people prefer, but Mill thinks "no intelligent human being would consent to be a fool, no instructed person would be an ignoramus, no person of feeling and conscience would be selfish and base."[10] Mill therefore concludes, famously, that "it is better to be a human being dissatisfied than a pig satisfied; better to be Socrates dissatisfied than a fool satisfied."[11] Rather than being a "pig philosophy," Mill thinks that utilitarianism respects the *dignity* of human life and the pursuit of happiness of all individuals.

Language and Expression

Mill wrote *Utilitarianism* for the general reader; as the British political philosopher Alan Ryan* puts it, his audience were not trained logicians.[12] Mill lays out the common objections to utilitarianism in a clear style and responds to each in turn. The text is engaging, and the nature of the debate and the ideas are clear. He also defines concepts and offers reasons in support of utilitarianism.

Utilitarianism has, nevertheless, been difficult to interpret. Mill's contemporaries, including the philosophers George Edward Moore* and Francis H. Bradley,* dismissed the text as inconsistent. Modern scholars such as Henry R. West* and David Lyons,* however, have offered a different picture. Mill's ideas are useful to students and academics in the fields of ethics and politics as he deals with concepts that remain relevant today, such as utilitarianism, security, justice, and liberty. Academic debates on the nature of morality and liberalism have frequently made reference to Mill's ideas, and many academic books have been devoted to analyzing his thought. Many people, including politicians and policy makers, take decisions based on utilitarian considerations, although it is difficult to tell whether they draw on Mill's specific ideas or on utilitarian thought in general.

---

1. John Stuart Mill, *Utilitarianism*, ed. Roger Crisp (Oxford: Oxford University Press, 1998), chapter 2.

2. Mill, *Utilitarianism*, 55.

3. Mill, *Utilitarianism*, 55.

4. Mill, *Utilitarianism*, 55.

5. Mill, *Utilitarianism*, 57.

6. Mill, *Utilitarianism*, 63.

7. Mill, *Utilitarianism*, 63–64.

8. Mill, *Utilitarianism*, 64.

9. Mill, *Utilitarianism*, 56.

10. Mill, *Utilitarianism*, 56–57.

11. Mill, Utilitarianism, 57.

12. Alan Ryan, *John Stuart Mill* (London: Routledge & Kegan Paul, 1974).

MODULE 6

SECONDARY IDEAS

KEY POINTS

* Mill argues that we are motivated to act according to utilitarian principles because of our conscience and our sympathy with others.

* Mill claims that justice and utilitarianism as a moral philosophy are entirely compatible and that security is of great importance for utilitarianism.

* Although the chapter on the relationship between justice and utility is the longest in the book, it has received relatively little attention in the literature.

Other Ideas

Having identified happiness as the goal of human life and defined the principle of utility* in *Utilitarianism*, John Stuart Mill went on to address a series of further concerns. Three themes in particular are worth singling out: motivation, justice, and security.

The first of these is our *motivation* to act in the morally right way. This is an important issue, because it addresses a tension in the utilitarian position; on the one hand, Mill says that our own happiness is the end (that is, the goal) of our life; it is what we all pursue. On the other hand, he claims that morally good actions often involve a sacrifice of one's own happiness for the sake of the general good. If the first claim is true, how could we be motivated to act on the basis of the second claim? Mill's answer is that our *conscience* and our *sympathy* toward mankind motivate us to do so. Or, as he puts it, we are motivated by "the social feelings of mankind; the desire to be in unity with our fellow creatures."[1]

The second theme is the concern that utilitarianism might conflict with justice. Utilitarianism is about maximizing the happiness of the greatest number of people. But critics argue that to achieve this, the interests of some people might sometimes have to be sacrificed for the sake of others. This would be unjust. For instance, an innocent person might have to be imprisoned in order to stop an angry crowd rioting; the imprisonment would lead to a greater overall happiness (it prevents a riot), but it would be an injustice for the innocent prisoner. Mill devotes his final chapter to proving that justice is vital for utilitarianism and does not conflict at all with the utilitarian view.

A third idea that Mill develops in *Utilitarianism* is that of security. Although *Utilitarianism* focuses mainly on happiness, Mill recognizes that a sense of security is crucial for people to feel happy. He introduces a political and legal element to the text, arguing that we should promote conditions that give people a sense of security in their lives.

> "The ultimate sanction, therefore, of all morality (external motives apart) being a subjective feeling in our own minds, I see nothing embarrassing to those whose standard is utility, in the question, what is the sanction of that particular standard? We may answer, the same as of all other moral standards: the conscientious feelings of mankind."
>
> ——John Stuart Mill, *Utilitarianism*

Exploring the Ideas

Mill argues that our conscience and our sympathy toward others motivate us to act morally; he does not believe that morality comes from external sanctions, such as laws, social punishment, or godly disapproval. The ultimate sanction of all morality, Mill claims, is "a subjective feeling in our own minds"—namely, "the conscientious feelings of mankind."[2] This is clever reasoning on Mill's behalf, as he can now claim that by helping others, we in fact serve our own happiness. He writes that "so long as [people] are co-operating, their ends are identified with those of others; there is at least a temporary feeling that the interests of others are their own interests."[3] When we sacrifice ourselves for the general happiness of others, we satisfy the desires of our conscience. This gives us intense pleasure and explains why we are motivated to act in such a way.

Mill argues that utilitarianism is not opposed to justice; rather, the idea of justice can be explained in terms of utility. He concludes that feelings about justice are based on the belief that people's behavior, social institutions, and social policies should not undermine our own happiness or the happiness of anyone else. In Mill's words:"The sentiment of justice appears to me to be, the animal desire to repel or retaliate a hurt or damage to oneself, or to those with whom one sympathizes, widened so as to include all persons, by the human capacity of enlarged sympathy, and the human conception of intelligent self-interest."[4] Mill thinks what is just is not opposed to the principles of utilitarianism. The two are compatible and justice is *required* and *grounded* by utilitarianism.

This gives us a direct grip on our third theme—why Mill thinks security is so important. Mill regards security as something that no person can do without, and calls it "the most vital of all interests."[5] Although people want a wide range of things, they all have an interest in furthering security. Since security is a common interest, in that everyone shares it, it should be maximized for all. It is therefore in the interest of all of us to promote security since in the long run our own individual interests will also be looked after. When our security is in danger, we think in terms of injustices, though in the end the importance of moral security is explained by utility—the ultimate good.

Overlooked

John Stuart Mill's *Utilitarianism* is a short work that has been subjected to serious scrutiny. One would think that no part of it has been neglected. Most analysis, however, has focused on particular passages of chapters 2 and 4, rather than providing an interpretation of the entire book. The final chapter that examines the relation between justice and utilitarianism has received far less attention than chapter 4, which attempts to prove that the principle of utility is true.

This could be due to the fact that early responses to *Utilitarianism* concentrated on Mill's proof of the principle of utility. Such thinkers, including the British philosophers Francis H. Bradley* and George Edward Moore,* tried to prove that Mill had committed a number of logical fallacies.* Modern scholars such as Henry R. West,* David Lyons,* Jonathan Riley,* and John Skorupski* have continued to draw on these earlier objections to Mill's proof of the principle of utility. As a result, the chapter on proofs has been the most popular among scholars.

---

1. John Stuart Mill, *Utilitarianism*, ed. Roger Crisp (Oxford: Oxford University Press, 1998), 77.
2. Mill, *Utilitarianism*, 75.
3. Mill, *Utilitarianism*, 78.
4. Mill, *Utilitarianism,* 97.
5. Mill, *Utilitarianism*, 98.

ACHIEVEMENT

KEY POINTS

- *Utilitarianism* develops the utilitarian views of the social reformer and philosopher Jeremy Bentham,* clarifies what utilitarianism is, and defends it against common criticisms.

- The text developed utilitarianism into one of the most important positions in moral philosophy and is ranked among the greatest works in moral philosophy.

- The text appears to contain logical fallacies,* however: it fails to clarify how higher and lower pleasures can affect calculations of overall happiness, and it is not clear whether Mill supported "rule utilitarianism"* or "act utilitarianism"* (two different ways of understanding what might define a "right" action).

Assessing the Argument

John Stuart Mill's main aim in *Utilitarianism* is to clarify utilitarianism as a position in moral philosophy, to develop the utilitarian views of the thinker Jeremy Bentham, and to defend his own version of utilitarianism against its critics. In this respect, the book has been a great success. Mill stated more clearly than anyone before him what utilitarianism is, and his distinction between higher and lower pleasures has had a lasting influence on debates in moral philosophy. His responses to critics of utilitarianism are detailed and full of insight. His proof of utilitarianism offered in chapter 4, although disputed, continues to attract attention. It has helped make utilitarianism the most widely supported view in moral philosophy today.

Utilitarianism is both one of the most important books in moral philosophy and the most widely read statement of utilitarian ideas. It is widely used in colleges and universities to introduce students to moral philosophy or ethics. The text's core arguments have been subject to numerous interpretations, even if no consensus has yet been reached on some issues. Scholars continue to debate the proper interpretation of Mill's arguments and his philosophical position.

> "Utilitarianism *is one of the most significant works in moral philosophy, ranking in importance alongside Aristotle's* Nicomachean Ethics and Immanuel Kant's* Groundwork of the Metaphysic of Morals.*"
>
> —— Roger Crisp, *Routledge Philosophy Guidebook to Mill on Utilitarianism*

Achievement in Context

When Mill wrote his text, utilitarianism was in decline as a respectable view in moral philosophy. In fact, Bentham's utilitarianism had faced serious criticisms and was viewed with suspicion by the majority of intellectuals of his time. Mill's text played a key role in strengthening the utilitarian position, to the point that it became a major force not only in philosophy but also in everyday politics.

Mill attacked ethical intuitionism*—the notion, very roughly, that it was possible to arrive at an understanding of ethical behavior through intuition alone—and said that the principle of utility*, not the will of God, was the foundation of morals. He also modified Bentham's version of utilitarianism. It is therefore not surprising that he was criticized by a wide range of thinkers, particularly intuitionists, fellow utilitarians, and religious scholars.

Despite these criticisms, his work was considered the best statement of utilitarianism available.[1] When the Irish historian and political theorist William E. H. Lecky* controversially failed to mention Mill in his *History of European Morals* (1869), even opponents of utilitarianism criticized him. They felt that Mill's utilitarianism was the best version of the doctrine and, as such, deserved attention.[2] According to the American scholar Jerome B. Schneewind,* the controversy marked the "point at which it became generally clear that to deal with utilitarianism one had to deal with Mill's version of it."[3]

Limitations

One limitation of Mill's *Utilitarianism* is that it seems to rely on a number of fallacies.* To go from his claim that happiness is the greatest good to the idea that we should therefore promote the *general* good has been viewed as a false inference. The fallacy he has been accused of is the "fallacy of composition,"* which arises when one infers that what is true of a part or parts is true of the whole ("all the parts of a bicycle are light; therefore a bicycle is light," for example). So although it may be true that *personal* happiness is our own ultimate good, it does not follow that *general* happiness is the ultimate good when we take all people together.

A second limitation is that Mill did not say how the difference between higher and lower pleasures should be calculated, other than stating that higher pleasures should have greater weight. According to Bentham's version of utilitarianism, every pleasure has equal value, and each individual unit of pleasure (sometimes called "utils") can be counted and added together. Precisely how that is meant to work in Mill's view of utilitarianism remains unclear.

A third limitation is that Mill defends a combination of what are now known as "rule utilitarianism" and "act utilitarianism." According to act utilitarianism, the moral goodness of an action depends on the degree to which *that action* affects the total amount of happiness. According to rule utilitarianism, by contrast, the moral goodness of an action depends on whether it fits with *rules* that generally positively affect the total amount of happiness. This is an important difference, as it can lead to contradictory claims about whether an action is right or not. Killing, for instance, can sometimes greatly increase the totality of happiness, although killing in general will not. Mill is not clear about which version of utilitarianism he supports. In defining utilitarianism early in the text, he writes that "*actions* are right in proportion as they tend to promote happiness, wrong as they tend to produce the reverse of happiness."[4] But a few pages later he writes that "the end [that is, the goal] of human action, is necessarily also the standard of morality; which may accordingly be defined, *the rules and precepts for human conduct*, by the observance of which an existence such as has been described might be, to the greatest extent possible, secured to all mankind."[5] In recent years, the British scholar of ethics Roger Crisp* has favored an interpretation of Mill along the lines of act utilitarianism.[6] Others have preferred the elements highlighted by rule utilitarianism.[7] And some think that we can find both elements in Mill's text.[8]

1. J. B. Schneewind, "Concerning Some Criticisms of Mill's *Utilitarianism*, 1861–76," in *James and John Stuart Mill: Papers of the Centenary Conference*, eds. John M. Robson and Michael Laine (Toronto and Buffalo: University of Toronto Press, 1976).

2. Schneewind, "Concerning Some Criticisms," 40.

3. Schneewind, "Concerning Some Criticisms," 40.

4. John Stuart Mill, *Utilitarianism*, ed. Roger Crisp (Oxford: Oxford University Press, 1998), 55.

5. Mill, *Utilitarianism*, 59.

6. See Roger Crisp, *Routledge Philosophy Guidebook to Mill on Utilitarianism* (London: Routledge, 1977), 102.

7. See Dale E. Miller, *John Stuart Mill: Moral, Social and Political Thought* (Cambridge: Cambridge University Press, 2010).

8. See Henry R. West, *An Introduction to Mill's Ethics* (Cambridge: Cambridge University Press, 2004).

MODULE 8
PLACE IN THE AUTHOR'S WORK

KEY POINTS

* Mill's moral philosophy is presented most clearly and systematically in *Utilitarianism*. But many of these ideas were prepared in earlier essays and publications by Mill.

* Mill's work shows great consistency of thinking over his lifetime, revealing a clear commitment to individual liberty and the increase of happiness for all.

* *Utilitarianism* is the most important of all his works and is as relevant today as it was when it was first published.

Positioning

John Stuart Mill started writing at an early age, contributing sophisticated articles in newspapers and periodicals before he was twenty. By the time *Utilitarianism* was published in 1861, he had already produced a number of essays and books and was regarded as one of the leading philosophers in Britain. His first major published text was *A System of Logic, Ratiocinative and Inductive* (1843), which argued in favor of logic as a method of proof and also targeted intuitionists.* In 1848, Mill published *Principles of Political Economy*, which became a major text in economic theory.

Mill's essay "On Liberty" (1859) offers a defense of individual liberty and argues that conduct that causes harm to others should be restricted. He returns to this theme in chapter 5 of *Utilitarianism*. Mill says it is entirely appropriate for society to interfere with individual behavior to prevent harm to others. In *Considerations on Representative Government* (1861) he argues for a form of government in Britain that is both representative and has greater public involvement. He says that public voting "ensures that opinion will be mobilized to both preserve and induce the public good."[1] In *Subjection of Women* (1869), he argues for the emancipation* of women and equality between the sexes. These texts show Mill's vision of a liberal society consisting of two parts:"creative individuality and the unity of will needed to sustain society."[2] They also reveal the development of a liberal doctrine grounded in utilitarianism. As the scholar Roger Crisp* concludes in his overview of Mill's work, "many of the issues discussed in *Utilitarianism* were foreshadowed in earlier essays by Mill."[3]

> "[Many] of the issues discussed in Utilitarianism were foreshadowed in earlier essays by Mill: the foundation of ethics and ethical understanding, the importance of first and secondary principles, the proof of utilitarianism, the sources of human happiness, moral motivation and the 'sanctions' of morality."
>
> —— Roger Crisp, Routledge *Philosophy Guidebook to Mill on Utilitarianism*

Integration

Despite the fact that Mill wrote on a range of different topics, his work demonstrates remarkable unity and consistency. In his *System of Logic*, written nearly 20 years before *Utilitarianism*,

Mill attacked intuitionists. He argued that if one is an intuitionist about ethics, then one should logically also be an intuitionist about science and accept that we can discover laws of nature without the help of observation. He considered this to be an absurd position. His attack on intuitionism* in *Utilitarianism* was clearly foreshadowed in his earlier work, as part of a long-term effort to promote empiricism* in all areas of human knowledge.

In *On Liberty* (1959) Mill offers a defense of individual liberty: people, he argues, should be allowed to act as they see fit, the only exception being if they intend to harm others. The latter qualification is the so-called harm principle, which states that "the only purpose for which power can be rightfully exercised over any member of a civilized community, against his will, is to prevent harm to others."[4] These are two themes that surface again in *Utilitarianism*, where Mill argues, first, that moral rightness consists in maximizing pleasure and minimizing pain and suffering. This means that people should not be impeded unless they do indeed harm others and so reduce the total sum of happiness. Second, Mill commits to the view that every person is of equal worth and that the happiness of each individual matters. One group cannot use others as a mere instrument to their own benefit. Maximizing the aggregate of happiness implies that each person has an equal claim to happiness.

Finally, in his *Subjection of Women* (1869), Mill extends his arguments to imply that women should have equal rights with men. Again, the maximization of the sum of all happiness is the main goal that inspires Mill. In his view this should include women as much as men; hence his passionate plea for the emancipation of women in nineteenth-century Britain.

Significance

The two best-known works of John Stuart Mill are *On Liberty* and *Utilitarianism*. *Utilitarianism* offers the most systematic and precise presentation of Mill's moral philosophy, drawing together a number of themes that had been in preparation since his early youth. Roger Crisp suggests that "in so far as Mill was an evangelist, *Utilitarianism*, first published as a series of three essays in 1861, can be seen as his bible."[5]

Mill was already well known as one of the most significant philosophers in Britain during his lifetime. *Utilitarianism* was not written for a small intellectual elite, however. It was published as a series of three essays in 1861 in *Fraser's Magazine for Town and Country*, a general and literary periodical that was founded in London in 1830 and aimed at a middle-class readership. Its first publication as a book was in 1863, and it was only then that Mill added the chapter on utility and justice.

Mill's reputation has changed little over the last 150 years. He is still considered a giant in philosophy. Most university courses in politics, economics, and moral philosophy deal extensively with Mill's views, especially those expressed in *Utilitarianism*. These views have lost none of their force or urgency, and it is unlikely that their significance will lessen any time soon.

1. Robert Devigne, *Reforming Liberalism: J. S. Mill's use of Ancient, Religious, Liberal, and Romantic Moralities* (New Haven, CT: Yale University Press, 2006), 218.

2. Devigne, *Routledge Philosophy Reforming Liberalism*, 219.

3. See Roger Crisp, *Routledge Philosophy Guidebook to Mill on Utilitarianism*. (London: Routledge, 1977), 12.

4. John Stuart Mill, *On Liberty* (London: Penguin, 2010), 21.

5. Crisp, *Routledge Philosophy Guidebook to Mill*, 7.

SECTION 3
IMPACT

THE FIRST RESPONSES

KEY POINTS

- Most of the early responses to Mill's *Utilitarianism* were hostile. He was criticized for making fallacious—that is, badly reasoned— inferences and being disloyal to the utilitarian position.

- Mill did not revise his arguments in light of these criticisms. Instead he tried to clarify what it was he had actually argued.

- There was no consensus on Mill's disputed claims during his lifetime. Indeed, debates on the true interpretation of Mill's views continue to this day.

Criticism

The majority of responses to John Stuart Mill's *Utilitarianism* in the first 15 years after its publication were hostile. Critics focused on alleged logical flaws in his arguments.[1] Four issues in particular were raised—namely, "his views on the derivability of moral notions from non-moral ones; his ... 'proof' of the principle of utility;* his distinction between higher and lower pleasures; and his claim that the rules of common-sense morality may be taken as the middle axioms [that is, accepted truths] of a utilitarian ethic."[2] Early critics such as George Edward Moore* questioned whether Mill's proof of the principle of utility was a fallacy.*[3] Mill tried to prove the principle of utility with the following argument:

"The only proof capable of being given that an object is visible, is that people actually see it. The only proof that sound is audible, is that people hear it ... In like manner, I apprehend the sole evidence it is possible to produce that anything is desirable, is that people do actually desire it."[4]

Moore accused Mill of drawing a false analogy between "desirable" and "visible," objecting that "the fallacy in this step is so obvious, that it is quite wonderful how he failed to see it. The fact is that 'desirable' does not mean 'able to be desired' as 'visible' means 'able to be seen'. The desirable simply means what *ought* to be desired or *deserves* to be desired."[5]

The influential philosopher Thomas H. Green,* in turn, objects to the claim that pleasure is the only object of desire. He points out that the object of desire and the pleasure produced from satisfying that desire are not the same.[6] For example, when we are hungry we do not seek pleasure, but food; so in this case, pleasure should not be seen as the object of desire. Moore, Green, and the philosopher Francis H. Bradley* were among early critics of Mill's distinction between quantitative and qualitative pleasures (that is, pleasures measured by quantity and those measured by quality) and questioned his use of the doctrine of hedonism.*[7]

It is important to note that not everyone objected to Mill's defense of utilitarianism. The Scottish philosopher James Seth,* for example, vigorously defended Mill's position in an article published in 1908.[8]

> "[When] I said that the general happiness is a good to the aggregate of all persons I did not mean that every human being's happiness is a good to every other human being; though I think in a good state of society and education it would be so."
>
> —— John Stuart Mill, *The Letters of John Stuart Mill*

Responses

Utilitarianism had been through four editions by 1871, with relatively minor revisions and additions. The philosopher Alexander Bain,* in his 1882 work entitled *John Stuart Mill*, said that he was "not aware that any change was made in reprinting [the first edition of *Utilitarianism*] as a volume, notwithstanding that it had its full share of hostile criticism as it came out in *Fraser*."[9] This implies that Mill did not respond to initial criticisms of his work in later editions. The American scholar Jerome Schneewind,* however, takes issue with the level of attention the work initially received. He claims that for the first seven or eight years *Utilitarianism* did not receive much attention in print, even though philosophical responses to the text were largely negative.[10]

Mill's argument that "happiness is a good: that each person's happiness is a good to that person, and the general happiness, therefore, a good to the aggregate of all persons"[11] has been the cause of much debate and has led to the charge of being a fallacy of composition.* Mill was aware of this criticism, and in a letter to a correspondent dated June 13, 1868, he makes his position clear: "I merely meant ... to argue that since A's happiness is a good, B's a good, C's a good, etc., the sum of these must be a good."[12]

Mill wrote the letter before the publication of Francis Bradley's critical essay published in his *Ethical Studies*. But Bradley failed to respond to it, instead rejecting the proof as Mill had stated it in *Utilitarianism*. It should be pointed out, though, that in Mill's letter he still proposes a general principle that if one puts good things together, one will end up with a good whole; Bradley's critique therefore still seems fair.[13]

Conflict and Consensus

Mill died in 1873, and after his death his opponents unleashed a wholesale attack on his version of utilitarianism. In the years that followed, a huge range of texts was produced both supporting and opposing those objections. The letters Mill wrote toward the end of his life suggest that he had not changed his views, believing that most of his critics had simply failed to understand him. Critics such as Bradley, in turn, paid no attention to the clarifications that Mill had added to his text. Instead they continued to point out what they thought was wrong in Mill's published version of *Utilitarianism*.

The debate over the correct interpretation of Mill's core ideas continues today. It is largely a repeat of earlier responses to Mill's core arguments. However, a number of modern scholars, among them Henry R. West* and David Lyons,* have provided a revisionist interpretation of Mill's utilitarianism—that is, an interpretation that questions orthodox opinion—arguing that his critics misread him. But there is a risk that in defending Mill against past and present critics, an inaccurate view is attributed to him. The conflict over Mill's defense of utilitarianism is nevertheless bound to continue for some time.

---

1. J. B. Schneewind, "Concerning Some Criticisms of Mill's *Utilitarianism*, 1861–76," in *James and*

John Stuart Mill: Papers of the Centenary Conference, eds. John M. Robson and Michael Laine (Toronto and Buffalo: University of Toronto Press, 1976).

2. Schneewind, "Concerning Some Criticisms," 41.

3. George Edward Moore, *Principia Ethica* (Cambridge: Cambridge University Press, 1903).

4. John Stuart Mill, *Utilitarianism*, ed. Roger Crisp (Oxford: Oxford University Press, 1998), 81.

5. Moore, *Principia Ethica*, 67.

6. Thomas Hill Green, *Prolegomena to Ethics* (Oxford: Clarendon Press, 2003).

7. See Moore, *Principia Ethica*; Green, *Prolegomena to Ethics*; Francis Herbert Bradley, *Ethical Studies* (London: Oxford University Press, 1962).

8. James Seth, "The Alleged Fallacies in Mill's 'Utilitarianism'," *The Philosophical Review* 17, no. 5 (1908): 469–488.

9. Cited in Schneewind, "Concerning Some Criticisms," 38.

10. Schneewind, "Concerning Some Criticisms."

11. Mill, *Utilitarianism*, 81.

12. John Stuart Mill, *The Letters of John Stuart Mill,* vol. 2, ed. with an introduction by Hugh S. R. Elliot, and a note on Mill's private life by Mary Taylor (New York: Longmans, Green & Co), 116.

13. Bradley, *Ethical Studies*, 113, n.1.

MODULE 10
THE EVOLVING DEBATE

KEY POINTS

- In the second half of the twentieth century utilitarianism became a major force in moral philosophy. It was further developed—but heavily criticized, too.

- In recent decades, many new schools of utilitarianism have been formed, including "two-level utilitarianism,"* "rule utilitarianism,"* and "preference utilitarianism."*

- Scholarship on utilitarianism is now common in a variety of academic disciplines, in both theoretical and applied form.

Uses and Problems

At the time of John Stuart Mill's death in 1873, empiricism,* which had dominated British philosophical thought and shaped *Utilitarianism*, had lost its position, first to idealism* (the philosophical position, very roughly, that "reality" is primarily a question of perception), and later to logical positivism* (the philosophical position that to be cognitively meaningful, a statement must be empirically verified or logically inferred from empirically verifiable statements). Little attention was paid to Mill's social philosophy until the second half of the twentieth century, when scholars such as Carl Wellman,* who specializes in ethics and philosophy of law, began to offer responses to the traditional criticisms of Mill's "proof" of the principle of utility.* Since then, commentators such as Henry R. West,* David Lyons,* and Jonathan Riley*[1] have offered a revisionist interpretation of

Mill's work, including explanations of how the principle of utility can be proven.

In the latter half of the twentieth century utilitarianism again became a major force in moral philosophy, with Mill's work being of central interest. Philosophers have expanded his version of utilitarianism in recent decades, forming new versions, finding new arguments in support of it, and introducing refinements.

At the same time, new criticisms of utilitarianism have also appeared. Three of these have become especially popular in the literature. The first is the idea of the "pleasure machine," introduced by the influential political philosopher Robert Nozick.* Nozick argues that if we were given the possibility to be hooked up to a machine that kept us permanently in a state of pleasure, most people would not describe this as a good life, despite the quantity of pleasure.[2] The second is the British thinker Philippa Foot's* famous trolley problem: pulling a lever so that an oncoming train kills one person instead of five seems morally different from pushing one person in front of the train to save five people's lives. Although the consequences are identical (one dead, five saved), the thought experiment tries to show that it is not just the results of our actions that matter in moral evaluations (as Mill says), but also our intentions.[3] And third, the British moral philosopher Bernard Williams* has also offered thought experiments intended to show that it can in fact be extremely immoral to act upon the greatest happiness principle. All these experiments question whether the principle of utility can be the foundation of all morality.[4]

"It is absurd to demand of ... a man, when the sums come in from the utility network which the projects of others have in part determined, that he should just step aside from his own project and decision and acknowledge the decision which utilitarian calculation requires."

———J. J. C. Smart and Bernard Williams, *Utilitarianism: For and Against*

Schools of Thought

Since the 1960s a number of different schools in utilitarianism have evolved. Here we shall focus on three such schools: rule utilitarianism,* two-level utilitarianism,* and preference utilitarianism.*

In 1953 the British philosopher James Opie Urmson* published an article arguing that Mill justifies moral rules ("Do not steal" and "Do not kill," for example) on the basis of the principle of utility.[5] The school of "rule utilitarianism"*—founded on the view that the moral goodness of an act depends on whether it obeys *rules* that generally increase the total amount of happiness— emerged after this. This view was developed to make it easier to calculate how much happiness or unhappiness an action might bring about. It is a view that, in its turn, has also received a great deal of criticism.

To overcome the limitations of rule utilitarianism, the British philosopher R. M. Hare* developed a view called "two-level utilitarianism." According to this school, although we should follow rules based on the principle of utility, in exceptional circumstances we may perform actions that break those rules. In these exceptional cases we are to take the position of act utilitarianism,* and assess (regardless of general rules) whether an action promotes happiness or not.[6] Two-level utilitarianism therefore combines rule and act utilitarianism.

The third important school is "preference utilitarianism," first developed by the Nobel prize-winning economist John Harsanyi,* but mainly associated with the Australian moral philosopher Peter Singer.*[7] This school argues that we are not in fact motivated to maximize pleasure and to minimize pain, but instead we are driven mainly to satisfy our own preferences. Like Mill, preference utilitarians claim that good actions promote good consequences. But unlike Mill, they believe that consequences will be good when they allow people to satisfy their own preferences.

In Current Scholarship

The importance of utilitarianism in current scholarship is hard to overstate. Utilitarianism is a major force in philosophy, psychology, economy, and politics. Virtually everyone dealing with normative* questions (questions concerning what is considered to be the normal or correct way of doing something), whether practical or theoretical, has to take a position on whether utilitarianism offers a suitable guide to what the right course of action is. Some scholars have focused on understanding precisely what thinkers like Jeremy Bentham* and Mill meant in their definitions, assertions, and arguments. Others point out what is wrong either with utilitarianism in general, or with specific versions of it. Still others are developing the utilitarian vision further and, like Mill, continue to defend it against its critics.

Utilitarianism plays a major role in applied ethics. For instance, debates about animal welfare have been approached from a utilitarian point of view. Animals can experience pleasure and pain, so there is clearly an argument that overall happiness or unhappiness should include that of animals. There are also debates in bioethics, where utilitarianism can help decide, for instance, which patients in health care should get public money. Areas such as poverty have also been approached from a utilitarian perspective. Utilitarianism is therefore developed—and also criticized—not just on a theoretical level, but also in a wide variety of applied forms.

---

1. See Jonathan Riley, "Mill's Extraordinary Moral Theory," *Politics, Philosophy and Economics* 9, no. 1 (2010): 67–116.

2. See Robert Nozick, *Anarchy, State, and Utopia* (New York: Basic Books, 1974), 42–45.

3. See Philippa Foot, "The Problem of Abortion and the Doctrine of the Double Effect," in *Virtues and Vices* (Oxford: Basil Blackwell, 1978).

4. See J. J. C. Smart and Bernard Williams, *Utilitarianism: For and Against* (Cambridge: Cambridge University Press, 1973).

5. J. O. Urmson, "The Interpretation of the Moral Philosophy of J. S. Mill," *The Philosophical Quarterly* 3 (1953): 33–39.

6. See R. M. Hare, *Proceedings of the Aristotelian Society, New Series* 73 (1972–73): 1–18.

7. See John Harsanyi, "Morality and the Theory of Rational Behaviour," in *Utilitarianism and Beyond*, eds. Amartya Sen and Bernard Williams (Cambridge: Cambridge University Press, 1982), 39–62; Peter Singer, *Practical Ethics* (Cambridge: Cambridge University Press, 1979).

IMPACT AND INFLUENCE TODAY

KEY POINTS

- *Utilitarianism* is viewed as a classic work in the history of philosophy. But it still inspires contemporary utilitarians who support their views with the insights offered by Mill.

- *Utilitarianism* is one of the three major traditions in moral philosophy—and Mill has powerful arguments why the other two might be reducible to utilitarianism.

- Philosophical debates about utilitarianism continue, with those over the "consequentializing" move becoming increasingly important.

Position

John Stuart Mill's *Utilitarianism* is both one of the most important books in moral philosophy and the most widely read account of utilitarianism. It has been widely used in colleges and universities to introduce students to moral philosophy or ethics. It is a classic text in the history of philosophy. Contemporary philosophers still refer to it, either to support their own strand of utilitarianism, or to criticize or defend aspects of utilitarianism that were already present in Mill's ideas. Modern utilitarians such as the Australian moral philosopher Peter Singer,* for instance, who believe they are continuing Mill's project of defending utilitarianism, continue to reference the text.

In debates about the nature of justice, *Utilitarianism* also continues to be an important source, mainly because of the fifth chapter, in which Mill lays out his view on the relationship between utility and justice. John Rawls,* the famous twentieth-century theoretician of justice, directly refers to *Utilitarianism* when criticizing the view that utility maximization is the foundation of justice.[1] Rawls considers it unjust to sacrifice the interests of a single individual in order to benefit others (according to utilitarianism, the morally right thing to do in some situations) unless the individual whose interests are sacrificed is compensated in some way. In developing this argument, Rawls engages directly with Mill's *Utilitarianism*. Mill's text is by no means an antiquated classic in the history of ideas. Despite the developments in the utilitarian tradition, it continues to hold a central place in contemporary moral philosophy.

> *"Current trends in philosophy make it easier to appreciate Mill, to rethink his work and put it to use, than it has been for a hundred years or more."*
>
> ——John Skorupski, "Introduction," in *Cambridge Companion to Mill*

Interaction

Utilitarianism is generally seen as one of the three major traditions in moral philosophy; Kantian deontology* (a moral philosophy centered on notions of duty) and Aristotelian virtue ethics* (a moral philosophy centered on the cultivation of virtuous behavior) are the two others. In *Utilitarianism*, Mill presents arguments against both traditions. He holds that when Kant says we should never act on a principle that cannot be universalized, "all he shows is that the *consequences* of their universal adoption would be such as no one would choose to incur."[2] Mill is more favorable toward virtue ethicists, especially with respect to their emphasis on education and development of good character. But again, he thinks that the reason *why* a good character is so important is best explained by the utility principle: good character will result in the performance of better activities. It will also lead to a greater amount of overall happiness, which is our main goal in life.

In his criticism of these traditions, Mill in fact turns virtue ethics and Kantian deontology into forms of utilitarianism. This type of attack on Kantian deontology and virtue ethics has been revived in recent years and labeled "consequentializing."[3] "To consequentialize," as one contributor to the debates writes, "is to take a putatively non-consequentialist moral theory and show that it is actually just another form of consequentialism."[4] Consequentializers agree with Mill that what ultimately matters is bringing about good consequences. Despite the difficulties that utilitarianism faces, in recent years this has proven to be a major challenge to schools in moral philosophy that focus on things other than consequences (intentions, for example, or character traits).

The Continuing Debate

The contemporary debates on utilitarianism take many forms. First, contemporary philosophers continue to identify which position classical utilitarians such as Jeremy Bentham* and John Stuart Mill actually defended. In addition to the contributions of Henry

R. West,* David Lyons,* Jonathan Riley,* and John Skorupski,* it is worth referring to a recent book by Fred Rosen of University College London titled *Classical Utilitarianism from Hume to Mill* (2003).[5] Second, philosophers continue to debate the validity of utilitarianism. The criticisms offered by Robert Nozick,* Philippa Foot,* and Bernard Williams* have sparked a great many responses from philosophers working today.

And third, the specific attack that consequentializers make on Kantian deontology and virtue ethics is debated among moral philosophers. Many Kantians and virtue ethicists have tried to come up with responses as to why their specific view cannot be "consequentialized." The professor of moral philosophy Christine Korsgaard* argues in a recent paper that neither Kant's nor Aristotle's position can be consequentialized, because the idea of a *general* good, one that is not good *for* someone, is incoherent.[6]

Although these debates are very recent and it will no doubt take some time before consensus is reached, it is worth pointing out that the "consequentializing move" has its roots in Mill's text. For these, as well as many other reasons, Skorupski, a professor of moral philosophy, claims that "current trends in philosophy make it easier to appreciate Mill, to rethink his work and put it to use, than it has been for a hundred years or more."[7]

---

1. John Rawls, *A Theory of Justice* (Cambridge, MA: Belknap Press of Harvard University, 1971).

2. John Stuart Mill, *Utilitarianism*, ed. Roger Crisp (Oxford: Oxford University Press, 1998), 52.

3. Campbell Brown, "Consequentialise This," *Ethics* 121, no. 4 (2011): 749–771.

4. Brown, "Consequentialise This," 749.

5. Fred Rosen, *Classical Utilitarianism from Hume to Mill* (London: Routledge, 2003).

6. Christine Korsgaard, "On Having a Good," *Philosophy: The Journal of the Royal Institute of Philosophy* 89, no. 3 (2014): 405–429.

7. John Skorpuski, "Introduction," in *Cambridge Companion to Mill*, ed. John Skorupski (Cambridge, Cambridge University Press, 1998), 2.

MODULE 12
WHERE NEXT?

KEY POINTS

* *Utilitarianism* will continue to be important, not only as an introductory text to utilitarianism, but also as a rich source for further developments of the theory.

* Contemporary philosophers continue to build on Mill's central arguments, trying to defend him against the criticisms that have been leveled against these arguments.

* *Utilitarianism* is a key text in moral philosophy because it is the first thorough and systematic explanation and defense of utilitarianism, supported by evidence.

Potential

We must assume that John Stuart Mill's *Utilitarianism* will continue to have an important influence in moral philosophy, economics, and political theory. As a clear and systematic presentation of utilitarianism, it will continue to serve as the founding text of one of the most important and popular theories in moral philosophy. The utilitarian tradition has grown extensively over the last decades. Several different schools of utilitarianism now argue over how to define the utility principle, in terms of acts, rules, or preferences, and so on. Although in recent years the hedonistic* (that is, pleasure-oriented) aspects of Mill's views have given way to the preference utilitarianism advanced by the notable Australian moral philosopher Peter Singer* and others, the text continues to have great appeal to modern moral philosophers.

Less attention has been paid to Mill in recent years in the literature on political liberalism. Liberal thinkers have focused instead on the highly influential twentieth-century moral philosopher John Rawls's* *A Theory of Justice*, which defined justice as fairness crucial for a liberal society to function. The publication of John Rawls's text in 1971 saw a growing interest in rights-based political philosophy in universities. At the time, utilitarianism dominated political philosophy; Rawls's text, however, revisited the idea of the "social contract"* between the government and the governed which utilitarianism had displaced. Hundreds of books and thousands of articles have been written in response to Rawls's work—evidence that rights-based political philosophy has by and large displaced the dominance of utilitarianism in political philosophy.

Since the 1950s, scholarship on Mill's *Utilitarianism* has focused on providing a correct interpretation of the text. The body of secondary literature on Mill's work is immense and continues to grow. It is likely that the text's core ideas will be even further developed, since many philosophers are now returning to Mill's ideas about human happiness, the principle of utility,* and the relationship between justice and utility. For the foreseeable future these will remain the main interests of philosophers actively engaging with the text.

> *"Since it is short, readable, polemical and eloquent, it has always offered an easy way into the complexities of moral philosophy and into the creed of the utilitarian movement."*
> ——Alan Ryan, "Introduction," *Utilitarianism and Other Essays*

Future Directions

Utilitarianism will continue to be developed, we must assume, often in direct dialogue with its fiercest critics. This mirrors the way Mill developed his own account in direct response to the critics of his day. Although the extent to which Mill's *Utilitarianism* will contribute to these developments is difficult to predict, several contemporary philosophers can be mentioned who show great promise in continuing to build on Mill's text.

The British philosopher Alan Ryan,* for instance, has subjected chapter 5 of the text to close analysis and combined the insights it contains with several chapters from Mill's *On Liberty* and sections of Mill's *A System of Logic*. Ryan concludes that the liberal doctrine that Mill developed is based on utilitarianism, thereby reestablishing a close connection between political liberalism and Mill's classical utilitarianism.[1] Scholars such as Ryan argue that Mill's *Utilitarianism* can better be understood if it is examined in relation to his other works.

The British political philosopher John Gray* and the philosophy scholar Fred Berger,* on the other hand, have examined Mill's central concept of happiness, and tried to show its significance to his understanding of rights, liberty, justice, freedom, and moral rules.[2] The scholar of ethics John Skorupski,* a long-standing admirer of Mill's work, is currently assessing the critique that John Rawls* leveled against Mill's concept of liberty. Ongoing research such as this might pave the way for a return of Mill's liberalism to the philosophical and political forefront.

These examples demonstrate that *Utilitarianism* is by no means an ornament in the history of outdated ideas. It continues to be a source of inspiration for moral philosophers who are utilitarians and a challenge for those who are not.

Summary

John Stuart Mill's *Utilitarianism* is a key work in the history of moral philosophy. It contains a systematic and thorough answer to the question, "How should we live our lives and how should we live together?"

Mill combines the insights of his father James Mill* and his tutor Jeremy Bentham,* together with ethical insights from ancient philosophers. He argues that happiness is the ultimate goal of human life, so that an action is right if it tends to increase the amount of human happiness and reduce the amount of unhappiness in the world. The appeal of Mill's account of utilitarianism stems from his effort to prove this principle of utility on an empirical* basis, by drawing a distinction between higher and lower pleasures, and by developing a utility-based view of justice.

Although Mill's work was heavily criticized at the time, it has had, and continues to have, a great impact on moral philosophy. Later developments of utilitarianism have built on Mill's insight and have often drawn inspiration from his text. Scholars also continue to investigate the central arguments of *Utilitarianism* in an attempt to show that the many criticisms it has faced are based on misinterpretations.

For anyone with an interest in questions about how to live a good life and how to act well in this world, *Utilitarianism* remains one of the most important texts to study.

---

1. Alan Ryan, "John Stuart Mill and the Art of Living," in *J. S. Mill's "On Liberty" in Focus*, eds. John Gray and G. W. Smith (London: Routledge, 1991); John Rees, *John Stuart Mill's "On Liberty"* (Oxford: Clarendon Press, 1985); Fred Berger, *Happiness, Justice and Freedom: The Moral and Political Philosophy of John Stuart Mill* (Berkeley: University of California Press, 1984).

2. John Gray, *Mill on Liberty: A Defence* (London: Routledge & Kegan Paul, 1983); Berger, *Happiness, Justice and Freedom*.

 GLOSSARY OF TERMS

1. **A priori:** the kind of knowledge that we can acquire independently of experience, such as mathematical knowledge.

2. **Act utilitarianism:** a moral theory stating that the moral goodness of an action depends on the degree to which *that action* affects the total amount of happiness.

3. **Calculative:** based on calculation.

4. **Deontology:** an approach to ethics founded on notions of duty and moral obligation that owes much to the eighteenth-century German philosopher Immanuel Kant.

5. **Emancipation:** being set free from legal, social, or political restrictions.

6. **Empirical:** based on experience, normally sensory experience.

7. **Empiricism:** the philosophical view that all knowledge is based on sense experience, as opposed to reason or intuition.

8. **Ethical intuitionism:** a doctrine that states that we have *a priori* knowledge of moral truths. Morality is known through intuition; intuition relates to direct awareness of certain objects.

9. **Ethical theory:** a reasoned account of how humans ought to behave or act.

10. **Fallacy:** a fault in reasoning that makes an argument or statement invalid.

11. **Fallacy of composition:** a false inference that arises when one infers that what is true of a part (or parts) is true of the whole. For example: all the parts of a bicycle are light, therefore a bicycle is light.

12. **French Revolution:** between 1789 and 1799, France experienced turmoil and a revolution that saw the end of the monarchy and the execution of King Louis XVI in 1793. Prior to the revolution, church leaders and the ruling classes wielded considerable power and led privileged lives while ordinary people faced poverty and high taxes.

13. **Greatest happiness principle:** the ethical *principle* that an action is right in so far as it promotes the *greatest happiness* of the *greatest* number of those

affected.

14. **Hedonism:** a doctrine that states that happiness or pleasure is the ultimate good.

15. **Hedonistic:** relying on the experience of pleasure.

16. **Idealism:** a philosophical doctrine that asserts that reality is made up of ideas, minds, or thoughts, not physical objects. Importantly, however, the term does not have a single meaning. In Britain, idealism was a dominant philosophical movement from the mid-nineteenth to the early twentieth century.

17. **Inductive school of ethics:** a school of philosophy that claims that moral knowledge is based on observation and experience.

18. **Industrial Revolution:** the period from the eighteenth to the nineteenth century when Britain experienced fundamental economic, social, and technological change as the nation moved to an economy based on industrial production.

19. **Intuitionism:** the belief that all knowledge is based on intuition—that is, that knowledge is acquired without appeal to reason, observation, or experience.

20. **Intuitive school of ethics:** a school of thought that points to the existence of moral intuition or sense. For such a school, general moral principles are known *a priori*; that is, they are self-evident.

21. **Jurist:** a legal scholar or legal theorist, studying theories of law.

22. **Kantian deontology:** a theory in moral philosophy that centers on duties and rules. Kant's version of deontology states that we must act for the sake of duty alone, by which he means that we must act in line with the moral law.

23. **Liberalism:** a political movement that defended individual freedom against unlimited state control.

24. **Logical fallacy:** a fault in reasoning that makes an argument or statement invalid.

25. **Logical positivism:** a theory of knowledge according to which cognitively

meaningful statements are those that can be empirically verified or logically inferred from empirically verifiable statements.

26. **Moral obligation:** something one must do or has a duty to do independent of personal or circumstantial factors; "Do not kill," for example.

27. **Normative:** relating to an ideal standard or model, or being based on what is considered to be the normal or correct way of doing something.

28. **Philosophical radicals:** a group of utilitarian thinkers who demanded social reform in England.

29. **Positivism:** the belief that scientific knowledge must be based directly or indirectly on sensory experience and tested via experiments.

30. **Preference utilitarianism:** a moral theory stating that the goodness of an action depends on whether it promotes people in being able to satisfy their own desires and preferences.

31. **Principle of utility:** this states that actions or behaviors are right in so far as they promote happiness or pleasure, or wrong if they tend to produce unhappiness or pain.

32. **Romanticism:** a nineteenth-century philosophical, artistic, and literary movement, prevalent in Germany, which valued non-rational aspects of human nature such as imagination and feelings.

33. **Rule utilitarianism:** a moral theory stating that the moral goodness of an action depends on whether it fits rules that normally result in a positive effect on the total amount of happiness.

34. **Social contract:** the idea that legitimate state authority must stem from the consent of the governed.

35. ***Tabula rasa***: a Latin phrase, often translated as "blank slate." In philosophy it is used to describe the idea that people are born without mental content and innate knowledge, and that all knowledge comes from experience and perception. The philosophical idea of *tabula rasa* originated from John Locke.

36. **Two-level utilitarianism:** a position in moral philosophy stating that we should follow rules based on the principle of utility, but in exceptional circumstances break those rules. These exceptional circumstances are when following a rule will result in a diminution of the total amount of happiness.

37. **Virtue ethics:** a label given to the ethical views of the Ancient Greeks, whereby the cultivation of virtues of excellences of character was deemed central.

1. **Aristippus (435–356 B.C.E.)** was an Ancient Greek ethical thinker who founded the Cyrenaic school of philosophy. A pupil of Socrates, he taught that the goal of life was to seek pleasure by maintaining control over adversity and prosperity.

2. **Aristotle (384–322 B.C.E.)** was a student of Plato and classical Greek philosopher whose ideas have shaped Western philosophy. His work covers a variety of subjects, including linguistics, physics, poetry, music, biology, politics, and ethics.

3. **Alexander Bain (1818–1903)** was a Scottish philosopher and educationalist who played a significant role in the development of modern psychology. He was also a radical follower of John Stuart Mill. One of his major works is *The Emotions and the Will* (1859).

4. **Cesare Beccaria (1738–1994)** was an Italian philosopher, politician, and jurist. He is best known for his work *On Crimes and Punishments* (1674), which was critical of the death penalty and torture.

5. **Jeremy Bentham (1748–1832)** is regarded as the father of modern utilitarianism. He was a social reformer, philosopher, and jurist who advocated a criminal justice system and state institutions that produce "the greatest happiness of the greatest number."

6. **Fred Berger (1937–1986)** was a professor of philosophy at the University of California, Davis. He wrote mainly on ethics and legal philosophy.

7. **Francis H. Bradley (1846–1924)** was a British philosopher who was best known for his work *Appearance and Reality: A Metaphysical Essay* (1893).

8. **Thomas Carlyle (1795–1881)** was a Scottish historian and philosopher, and one of the prominent literary figures of his time.

9. **Samuel Taylor Coleridge (1772–1834)** was a leading English Romantic poet.

10. **Auguste Comte (1798–1857)** was a French philosopher and social theorist who coined the term "sociology." He also developed the doctrine of positivism.

11. **Roger Crisp (b. 1961)** is Professor of Moral Philosophy and Uehiro Fellow and Tutor in Philosophy at St. Anne's College, Oxford. His work is primarily concerned with ethics.

12. **Epicurus (341–270 B.C.E.)** was a classical Greek philosopher known for advancing Epicureanism—a school of philosophy that advocated the pursuit of pleasure, particularly mental pleasure, which it considered the highest good.

13. **Philippa Foot (1920–2010)** was a British philosopher best known for her work in ethics, especially for her effort at revitalizing ancient virtue ethics. Her best-known works are *Virtues and Vices and Other Essays in Moral Philosophy* (1978) and *Natural Goodness* (2001).

14. **John Gray (b. 1948)** is an English political philosopher. The author of several best-selling works dealing with philosophical issues, he is noted for his work in analytic philosophy and the history of ideas.

15. **Thomas Hill Green (1836–1882)** was a distinguished and influential British philosopher and political theorist.

16. **Richard Mervyn Hare (1919–2002)** was a British philosopher at Oxford University specializing in metaethics. His best-known works are *The Language of Morals* (1952), *Freedom and Reason* (1963), and *Moral Thinking* (1981), in which he defends the view known as "Universal Prescriptivism."

17. **John Harsanyi (1920–2000)** was a Hungarian American economist and Nobel prize winner, best known for his work in game theory.

18. **Claude Adrien Helvétius (1715–1771)** was a French philosopher and philanthropist who advocated a materialist theory of human nature. This theory states that our actions are determined by our surroundings.

19. **David Hume (1711–1776)** was a Scottish philosopher, economist, and historian best known for his empiricism (the idea that all knowledge arises from the senses) and skepticism (the philosophical position that true knowledge is completely unattainable).

20. **Francis Hutcheson (1694–1746)** was an Irish-born Scottish philosopher

and Professor of Moral Philosophy at the University of Glasgow. He is best known for defending the moral sense theory and moral sentimentalism, a theory that states that humans have a moral sense through which they can approve or disapprove of human action.

21. **Immanuel Kant (1724–1804)** was a key eighteenth-century German philosopher who played a significant role in the development of modern philosophy. He is known for one of the most influential works in philosophy, the *Critique of Pure Reason* (1781), and for *Groundwork of the Metaphysics of Morals* (1785).

22. **Christine Korsgaard (b. 1952)** is a professor of philosophy at Harvard University, specializing in moral philosophy. She is best known for her work on Kant and Aristotle, and for her critique of utilitarianism.

23. **William E. H. Lecky (1838–1903)** was an Irish historian and political theorist, best known for his *A History of the Rise and Influence of Rationalism in Europe* (1865) and *A History of European Morals from Augustus to Charlemagne* (1869).

24. **John Locke (1632–1704)** was a British philosopher, medical researcher, and academic. He is known as the father of classical liberalism.

25. **David Lyons (b. 1935)** is a professor of law and professor of philosophy at Boston University, specializing in ethics and jurisprudence.

26. **Harriet Taylor Mill (1807–1858)** was a British philosopher and campaigner for women's rights who made significant contributions to the utilitarian movement.

27. **James Mill (1773–1836)** was a Scottish philosopher, historian, and economist who was prominent as a representative of the utilitarian school of thought. He wrote a number of articles on various subjects, including education, government, prisons, and colonies. One of his leading texts is *Elements of Political Economy* (1821). He was the father of John Stuart Mill.

28. **George Edward Moore (1873–1958)** was an influential British philosopher, best known for his *Principia Ethica*, first published in 1903.

29. **Robert Nozick (1938–2002)** was one of the most influential political philosophers of the late twentieth century, best known for his *Anarchy, State, and Utopia* (1974).

30. **Plato (427–347 B.C.E.)** was one of the best-known and most influential classical Greek philosophers. His text *The Republic*, which focuses on issues such as society, justice, and the individual, is considered to be one of the greatest philosophical texts ever written.

31. **John Rawls (1921–2002)** was an American political philosopher who was one of the leading thinkers in twentieth-century political philosophy and a strong advocate of individual rights. He is best known for his *A Theory of Justice* (1971), a landmark book of the twentieth century.

32. **David Ricardo (1772–1823)** was a British political economist whose writings made significant contributions to labor markets and international trade.

33. **Jonathan Riley** is a professor of philosophy and political economy at Tulane University, United States, specializing in utilitarianism, especially the philosophy of John Stuart Mill.

34. **John Ruskin (1819–1900)** is regarded as one of the greatest social commentators and art critics of the Victorian era.

35. **Alan Ryan (b. 1940)** is a British political philosopher and professor emeritus of political theory at the University of Oxford. He has written extensively on John Stuart Mill.

36. **Jerome B. Schneewind (b. 1930)** is an American academic and philosopher who is currently a professor emeritus of philosophy at Johns Hopkins University in Maryland.

37. **James Seth (1860–1925)** was a Scottish philosopher who held a chair in moral philosophy at Edinburgh University for 26 years.

38. **Peter Singer (b. 1946)** is an Australian moral philosopher affiliated with Princeton University and the University of Melbourne. He specializes in applied ethics, defending preference utilitarianism, and is best known for his *Animal Liberation* (1975) and his highly controversial support of infanticide.

39. **John Skorupski (b. 1946)** is a professor emeritus of moral philosophy at the University of St. Andrews, specializing in ethics, epistemology, and moral philosophy, and best known for his *The Domain of Reasons* (2010).

40. **Socrates (470–390 B.C.E.)** was the classical Greek philosopher who founded Western philosophy, and whose ideas were passed on to us primarily by his disciple Plato.

41. **James Opie Urmson (1915–2012)** was a philosopher and classicist who spent most of his professional career at the University of Oxford, specializing in ethics, ancient philosophy, and the work of George Berkeley.

42. **Carl Wellman** is a professor emeritus of philosophy, and Hortense and Tobias Lewin Distinguished University Professor in the Humanities at Washington University in St. Louis, specializing in ethics and philosophy of law.

43. **Henry R. West** is a professor emeritus of philosophy at Macalester College, Minnesota, specializing in utilitarianism, especially the philosophy of John Stuart Mill.

44. **William Whewell (1794–1866)** was a British philosopher, scientist, and theologian who was opposed to empiricism.

45. **Bernard Williams (1929–2003)** was a British moral philosopher at Cambridge University, and is considered to be one of the greatest British philosophers of the twentieth century.

46. **William Wordsworth (1770–1850)** was a leading English Romantic poet.

 **WORKS CITED**

1. Aristotle. *Nicomachean Ethics*. In *The Complete Works of Aristotle, The Revised Oxford Translation*, edited by Jonathan Barnes. Princeton, NJ: Princeton University Press, 1995.

2. Bentham, Jeremy. *An Introduction to the Principles of Morals and Legislation*. Edited by J. H. Burns and H. L. A. Hurt. London: Athlone, 1970.

3. Berger, Fred. *Happiness, Justice and Freedom: The Moral and Political Philosophy of John Stuart Mill*. Berkeley: University of California Press, 1984.

4. Bradley, Francis Herbert. *Ethical Studies*. London: Oxford University Press, 1962.

5. Brown, Campbell. "Consequentialise This." *Ethics* 121, no. 4 (2011): 749–771.

6. Crisp, Roger. "Introduction." In *Utilitarianism*, by John Stuart Mill, edited by Roger Crisp, 5–32. Oxford: Oxford University Press, 1998.

7. ———. *Routledge Philosophy Guidebook to Mill on Utilitarianism*. London: Routledge, 1977.

8. Devigne, Robert. *Reforming Liberalism: J. S. Mill's Use of Ancient, Religious, Liberal, and Romantic Moralities*. New Haven, CT: Yale University Press, 2006.

9. Foot, Philippa. "The Problem of Abortion and the Doctrine of the Double Effect." In *Virtues and Vices*. Oxford: Basil Blackwell, 1978.

10. Gray, John. *Mill on Liberty: A Defence*. London: Routledge & Kegal Paul, 1983.

11. Green, Thomas Hill. *Prolegomena to Ethics*. Edited by David O. Brink. Oxford: Clarendon Press, 2003.

12. Hare, R. M. "Principles." *Proceedings of the Aristotelian Society, New Series* 73 (1972–73): 1–18.

13. Harsanyi, John. "Morality and the Theory of Rational Behaviour." In *Utilitarianism and Beyond*, edited by Amartya Sen and Bernard Williams, 39–62. Cambridge: Cambridge University Press, 1982.

14. Kant, Immanuel. *Groundwork of the Metaphysics of Morals*. Edited and translated by Mary J. Gregor. Cambridge: Cambridge University Press, [1785] 1998.

15. Korsgaard, Christine. "On Having a Good." *Philosophy: The Journal of the Royal Institute of Philosophy* 89, no. 3 (2014): 405–429.

16. Mill, John Stuart. "Bentham." In *Utilitarianism and Other Essays*, edited by Alan Ryan, 132–176. London: Penguin Books, 2004.

17. ———. "Coleridge." In *Utilitarianism and Other Essays*, edited by Alan Ryan, 177–226. London: Penguin Books, 2004.

18. ———. *A System of Logic, Ratiocinative and Inductive: Being a Connected View of the Principles of Evidence and the Methods of Scientific Investigation*. Edited by J. M. Robson. Toronto: University of Toronto Press, 1973.

19. ———. *Autobiography*. Edited with an introduction and notes by Jack Stillinger. London: Oxford University Press, 1971.

20. ———. *Considerations on Representative Government*. London: Longmans, Green & Co., 1888.

21. ———. *On Liberty*. London: Penguin, 2010.

22. ———. *Principles of Political Economy: With Some of their Applications to Social Philosophy*. London: Longmans, Green & Co., 1911.

23. ———. *The Letters of John Stuart Mill*, vol. 2. Edited with an introduction by Hugh S. R. Elliot, with a note on Mill's private life by Mary Taylor. New York: Longmans, Green & Co., 1910.

24. ———. *The Subjection of Women*. London: Longmans, Green & Co., 1883.

25. ———. *Utilitarianism*. Edited by Roger Crisp. Oxford: Oxford University Press, 1998.

26. Miller, Dale E., *John Stuart Mill: Moral, Social and Political Thought*. Cambridge: Cambridge University Press, 2010.

27. Moore, G. E. *Principia Ethica*. Cambridge: Cambridge University Press, 1993.

28. Nozick, Robert. *Anarchy, State, and Utopia*. Oxford: Blackwell, 1974.

29. Rawls, John. *A Theory of Justice*. Delhi: Belknap Press of Harvard University, 1971.

30. Rees, John. *John Stuart Mill's "On Liberty"*. Oxford: Clarendon Press, 1985.

31. Riley, Jonathan. "Mill's Extraordinary Moral Theory." *Politics, Philosophy and Economics* 9, no. 1 (2010): 67–116.

32. Rosen, Fred. *Classical Utilitarianism from Hume to Mill*. London: Routledge,

2003.

33. Ryan, Alan. "Introduction." In *Utilitarianism and Other Essays*, by John Stuart Mill, edited by Alan Ryan, 7–63. London: Penguin Books, 2004.

34. ———. *John Stuart Mill*. London: Routledge & Kegan Paul, 1974.

35. ———. "John Stuart Mill and the Art of Living." In *J. S. Mill's "On Liberty" in Focus*, edited by John Gray and G. W. Smith, 162–168. London: Routledge, 1991.

36. ———. *On Politics: A History of Political Thought from Herodotus to the Present*. London: Allen Lane, 2012.

37. Schneewind, J. B. "Concerning Some Criticisms of Mill's *Utilitarianism*, 1861–76." In *James and John Stuart Mill: Papers of the Centenary Conference*, edited by John M. Robson and Michael Laine, 35–54. Toronto and Buffalo: University of Toronto Press, 1976.

38. Seth, James. "The Alleged Fallacies in Mill's 'Utilitarianism'." *The Philosophical Review* 17, no. 5 (1908): 469–488.

39. Singer, Peter. *Practical Ethics*. Cambridge: Cambridge University Press, 1979.

40. Skorpuski, John. "Introduction." In *Cambridge Companion to Mill*, edited by John Skorupski, 1–34. Cambridge, Cambridge University Press, 1998.

41. Smart, J. J. C. and Bernard Williams. *Utilitarianism: For and Against*. Cambridge: Cambridge University Press, 1973.

42. Urmson, J. O. "The Interpretation of the Moral Philosophy of J. S. Mill." *The Philosophical Quarterly* 3 (1953): 33–39.

43. West, Henry R. *An Introduction to Mill's Ethics*. Cambridge: Cambridge University Press, 2004.

原书作者简介

苏格兰哲学家詹姆斯·穆勒之子、哲学家和经济学家约翰·斯图亚特·穆勒幼时是个神童，3 岁学习希腊语，8 岁学习拉丁语，13 岁学习政治经济学。不难预料，他在 20 岁时精神崩溃了。但他得以康复，继续为东印度公司——英国在印度的代理政府工作，并在此后成为一名自由党议员。穆勒是最早要求给予妇女投票权的先驱之一。他支持了若干社会改革，并在自己的著作中帮助定义了"自由"的概念。1873 年穆勒于法国去世。

本书作者简介

帕特里克·汤姆拥有圣母大学、利兹大学和津巴布韦大学硕士学位，是圣安德鲁斯大学政治学与国际关系博士，现工作于津巴布韦政策对话研究所。

桑德尔·韦尔克霍芬是华威大学哲学博士，现于乌得勒支大学哲学与宗教研究学部从事伦理学和医学哲学的研究。

世界名著中的批判性思维

《世界思想宝库钥匙丛书》致力于深入浅出地阐释全世界著名思想家的观点，不论是谁、在何处都能了解到，从而推进批判性思维发展。

《世界思想宝库钥匙丛书》与世界顶尖大学的一流学者合作，为一系列学科中最有影响的著作推出新的分析文本，介绍其观点和影响。在这一不断扩展的系列中，每种选入的著作都代表了历经时间考验的思想典范。通过为这些著作提供必要背景、揭示原作者的学术渊源以及说明这些著作所产生的影响，本系列图书希望让读者以新视角看待这些划时代的经典之作。读者应学会思考、运用并挑战这些著作中的观点，而不是简单接受它们。

ABOUT THE AUTHOR OF THE ORIGINAL WORK

Son of the Scottish philosopher **James Mill**, the philosopher and economist John Stuart Mill was a child prodigy who learned Greek at three, Latin at eight, and was studying political economy at 13. Unsurprisingly, he suffered a nervous breakdown at 20, but recovered to continue working for the East India Company—Britain's proxy government in India—and later become a Liberal Member of Parliament. Mill was one of the first people to demand the right to vote for women. He backed a number of social reforms, and helped define concepts of freedom in his writing. He died in 1873 in France.

ABOUT THE AUTHORS OF THE ANALYSIS

Dr Patrick Tom holds masters degrees from Notre Dame, Leeds and the University of Zimbabwe, and a PhD in politics and international relations from the University of St Andrews. He currently works for the Zimbabwe Policy Dialogue Institute.
Dr Sander Werkhoven holds a PhD in philosophy from the University of Warwick. He is currently a member of the Department of Philosophy and Religious Studies at the University of Utrecht, where he specialises in ethics and the philosophy of medicine.

ABOUT MACAT
GREAT WORKS FOR CRITICAL THINKING

Macat is focused on making the ideas of the world's great thinkers accessible and comprehensible to everybody, everywhere, in ways that promote the development of enhanced critical thinking skills.

It works with leading academics from the world's top universities to produce new analyses that focus on the ideas and the impact of the most influential works ever written across a wide variety of academic disciplines. Each of the works that sit at the heart of its growing library is an enduring example of great thinking. But by setting them in context — and looking at the influences that shaped their authors, as well as the responses they provoked — Macat encourages readers to look at these classics and game-changers with fresh eyes. Readers learn to think, engage and challenge their ideas, rather than simply accepting them.

批判性思维与《功利主义》

首要批判性思维技巧：理性化思维

次要批判性思维技巧：阐释

约翰·斯图亚特·穆勒 1861 年出版的《功利主义》一直是最负盛名和最有影响力的道德哲学著作之一。它也是批判性思维的典范——穆勒在书中运用理性化思维和阐释的技巧创建出结构严谨、滴水不漏、令人信服的论点，来论证他在伦理学核心问题上的立场。

对穆勒来说，关键是给出有关是非对错的有效定义，并由此推导出自己的道德理论。提出有效的、可供辩护的定义是良好的阐释思维的重要方面，而穆勒很早就提出了他的（定义）。他认为，行为如果增进了幸福，就是好的；如果减少了幸福，就是坏的。但是必须明确，重要的并非是我们自己的幸福，而是受该特定行为影响的所有人的幸福的总和。穆勒从对道德的阐释出发，考虑了幸福的不同种类和性质后，论证出一个系统的、清晰连贯的框架来计算和判断整体幸福。

穆勒的论证如同任何典型推理一样，总是考虑到可能的反对意见，并将它们安排进书的章节里，以便在开展论证时承认其存在并予以反驳。

CRITICAL THINKING AND *UTILITARIANISM*

• Primary critical thinking skill: REASONING
• Secondary critical thinking skill: INTERPRETATION

John Stuart Mill's 1861 *Utilitarianism* remains one of the most widely known and influential works of moral philosophy ever written. It is also a model of critical thinking—one in which Mill's reasoning and interpretation skills are used to create a well-structured, watertight, persuasive argument for his position on core questions in ethics.

The central question, for Mill, was to decide upon a valid definition of right and wrong, and reason out his moral theory from there. Laying down valid, defensible definitions is a crucial aspect of good interpretative thinking, and Mill gets his in as early as possible. Actions are good, he suggests, if they increase happiness, and bad if they reduce happiness. But, vitally, it is not our own happiness that matters, but the total happiness of all those affected by a given action. From this interpretation of moral good, Mill is able to systematically reason out a coherent framework for calculating and judging overall happiness, while considering different kinds and qualities of happiness.

Like any good example of reasoning, Mill's argument consistently takes account of possible objections, building them into the structure of the book in order to acknowledge and counter them as he goes.

《世界思想宝库钥匙丛书》简介

《世界思想宝库钥匙丛书》致力于为一系列在各领域产生重大影响的人文社科类经典著作提供独特的学术探讨。每一本读物都不仅仅是原经典著作的内容摘要，而是介绍并深入研究原经典著作的学术渊源、主要观点和历史影响。这一丛书的目的是提供一套学习资料，以促进读者掌握批判性思维，从而更全面、深刻地去理解重要思想。

每一本读物分为3个部分：学术渊源、学术思想和学术影响，每个部分下有4个小节。这些章节旨在从各个方面研究原经典著作及其反响。

由于独特的体例，每一本读物不但易于阅读，而且另有一项优点：所有读物的编排体例相同，读者在进行某个知识层面的调查或研究时可交叉参阅多本该丛书中的相关读物，从而开启跨领域研究的路径。

为了方便阅读，每本读物最后还列出了术语表和人名表（在书中则以星号＊标记），此外还有参考文献。

《世界思想宝库钥匙丛书》与剑桥大学合作，理清了批判性思维的要点，即如何通过6种技能来进行有效思考。其中3种技能让我们能够理解问题，另3种技能让我们有能力解决问题。这6种技能合称为"批判性思维PACIER模式"，它们是：

分析：了解如何建立一个观点；

评估：研究一个观点的优点和缺点；

阐释：对意义所产生的问题加以理解；

创造性思维：提出新的见解，发现新的联系；

解决问题：提出切实有效的解决办法；

理性化思维：创建有说服力的观点。

THE MACAT LIBRARY

The Macat Library is a series of unique academic explorations of seminal works in the humanities and social sciences — books and papers that have had a significant and widely recognised impact on their disciplines. It has been created to serve as much more than just a summary of what lies between the covers of a great book. It illuminates and explores the influences on, ideas of, and impact of that book. Our goal is to offer a learning resource that encourages critical thinking and fosters a better, deeper understanding of important ideas.

Each publication is divided into three Sections: Influences, Ideas, and Impact. Each Section has four Modules. These explore every important facet of the work, and the responses to it.

This Section-Module structure makes a Macat Library book easy to use, but it has another important feature. Because each Macat book is written to the same format, it is possible (and encouraged!) to cross-reference multiple Macat books along the same lines of inquiry or research. This allows the reader to open up interesting interdisciplinary pathways.

To further aid your reading, lists of glossary terms and people mentioned are included at the end of this book (these are indicated by an asterisk [*] throughout) — as well as a list of works cited.

Macat has worked with the University of Cambridge to identify the elements of critical thinking and understand the ways in which six different skills combine to enable effective thinking.

Three allow us to fully understand a problem; three more give us the tools to solve it. Together, these six skills make up the PACIER model of critical thinking. They are:

ANALYSIS — understanding how an argument is built
EVALUATION — exploring the strengths and weaknesses of an argument
INTERPRETATION — understanding issues of meaning
CREATIVE THINKING — coming up with new ideas and fresh connections
PROBLEM-SOLVING — producing strong solutions
REASONING — creating strong arguments

"《世界思想宝库钥匙丛书》提供了独一无二的跨学科学习和研究工具。它介绍那些革新了各自学科研究的经典著作，还邀请全世界一流专家和教育机构进行严谨的分析，为每位读者打开世界顶级教育的大门。"

—— 安德烈亚斯·施莱歇尔，
经济合作与发展组织教育与技能司司长

"《世界思想宝库钥匙丛书》直面大学教育的巨大挑战……他们组建了一支精干而活跃的学者队伍，来推出在研究广度上颇具新意的教学材料。"

—— 布罗尔斯教授、勋爵，剑桥大学前校长

"《世界思想宝库钥匙丛书》的愿景令人赞叹。它通过分析和阐释那些曾深刻影响人类思想以及社会、经济发展的经典文本，提供了新的学习方法。它推动批判性思维，这对于任何社会和经济体来说都是至关重要的。这就是未来的学习方法。"

—— 查尔斯·克拉克阁下，英国前教育大臣

"对于那些影响了各自领域的著作，《世界思想宝库钥匙丛书》能让人们立即了解到围绕那些著作展开的评论性言论，这让该系列图书成为在这些领域从事研究的师生们不可或缺的资源。"

—— 威廉·特朗佐教授，加利福尼亚大学圣地亚哥分校

"Macat offers an amazing first-of-its-kind tool for interdisciplinary learning and research. Its focus on works that transformed their disciplines and its rigorous approach, drawing on the world's leading experts and educational institutions, opens up a world-class education to anyone."

—— Andreas Schleicher, Director for Education and Skills, Organisation for Economic Co-operation and Development

"Macat is taking on some of the major challenges in university education... They have drawn together a strong team of active academics who are producing teaching materials that are novel in the breadth of their approach."

—— Prof Lord Broers, former Vice-Chancellor of the University of Cambridge

"The Macat vision is exceptionally exciting. It focuses upon new modes of learning which analyse and explain seminal texts which have profoundly influenced world thinking and so social and economic development. It promotes the kind of critical thinking which is essential for any society and economy. This is the learning of the future."

—— Rt Hon Charles Clarke, former UK Secretary of State for Education

"The Macat analyses provide immediate access to the critical conversation surrounding the books that have shaped their respective discipline, which will make them an invaluable resource to all of those, students and teachers, working in the field."

—— Prof William Tronzo, University of California at San Diego

TITLE	中文书名	类别
An Analysis of Arjun Appadurai's *Modernity at Large: Cultural Dimensions of Globalisation*	解析阿尔君·阿帕杜莱《消失的现代性：全球化的文化维度》	人类学
An Analysis of Claude Lévi-Strauss's *Structural Anthropology*	解析克劳德·列维–斯特劳斯《结构人类学》	人类学
An Analysis of Marcel Mauss's *The Gift*	解析马塞尔·莫斯《礼物》	人类学
An Analysis of Jared M. Diamond's *Guns, Germs, and Steel: The Fate of Human Societies*	解析贾雷德·戴蒙德《枪炮、病菌与钢铁：人类社会的命运》	人类学
An Analysis of Clifford Geertz's *The Interpretation of Cultures*	解析克利福德·格尔茨《文化的解释》	人类学
An Analysis of Philippe Ariès's *Centuries of Childhood: A Social History of Family Life*	解析菲力浦·阿利埃斯《儿童的世纪：旧制度下的儿童和家庭生活》	人类学
An Analysis of W. Chan Kim & Renée Mauborgne's *Blue Ocean Strategy*	解析金伟灿/勒妮·莫博涅《蓝海战略》	商业
An Analysis of John P. Kotter's *Leading Change*	解析约翰·P.科特《领导变革》	商业
An Analysis of Michael E. Porter's *Competitive Strategy: Techniques for Analyzing Industries and Competitors*	解析迈克尔·E.波特《竞争战略：分析产业和竞争对手的技术》	商业
An Analysis of Jean Lave & Etienne Wenger's *Situated Learning: Legitimate Peripheral Participation*	解析琼·莱夫/艾蒂纳·温格《情境学习：合法的边缘性参与》	商业
An Analysis of Douglas McGregor's *The Human Side of Enterprise*	解析道格拉斯·麦格雷戈《企业的人性面》	商业
An Analysis of Milton Friedman's *Capitalism and Freedom*	解析米尔顿·弗里德曼《资本主义与自由》	商业
An Analysis of Ludwig von Mises's *The Theory of Money and Credit*	解析路德维希·冯·米塞斯《货币和信用理论》	经济学
An Analysis of Adam Smith's *The Wealth of Nations*	解析亚当·斯密《国富论》	经济学
An Analysis of Thomas Piketty's *Capital in the Twenty-First Century*	解析托马斯·皮凯蒂《21世纪资本论》	经济学
An Analysis of Nassim Nicholas Taleb's *The Black Swan: The Impact of the Highly Improbable*	解析纳西姆·尼古拉斯·塔勒布《黑天鹅：如何应对不可预知的未来》	经济学
An Analysis of Ha-Joon Chang's *Kicking Away the Ladder*	解析张夏准《富国陷阱：发达国家为何踢开梯子》	经济学
An Analysis of Thomas Robert Malthus's *An Essay on the Principle of Population*	解析托马斯·马尔萨斯《人口论》	经济学

An Analysis of John Maynard Keynes's *The General Theory of Employment, Interest and Money*	解析约翰·梅纳德·凯恩斯《就业、利息和货币通论》	经济学
An Analysis of Milton Friedman's *The Role of Monetary Policy*	解析米尔顿·弗里德曼《货币政策的作用》	经济学
An Analysis of Burton G. Malkiel's *A Random Walk Down Wall Street*	解析伯顿·G. 马尔基尔《漫步华尔街》	经济学
An Analysis of Friedrich A. Hayek's *The Road to Serfdom*	解析弗里德里希·A. 哈耶克《通往奴役之路》	经济学
An Analysis of Charles P. Kindleberger's *Manias, Panics, and Crashes: A History of Financial Crises*	解析查尔斯·P. 金德尔伯格《疯狂、惊恐和崩溃：金融危机史》	经济学
An Analysis of Amartya Sen's *Development as Freedom*	解析阿马蒂亚·森《以自由看待发展》	经济学
An Analysis of Rachel Carson's *Silent Spring*	解析蕾切尔·卡森《寂静的春天》	地理学
An Analysis of Charles Darwin's *On the Origin of Species: by Means of Natural Selection, or The Preservation of Favoured Races in the Struggle for Life*	解析查尔斯·达尔文《物种起源》	地理学
An Analysis of World Commission on Environment and Development's *The Brundtland Report, Our Common Future*	解析世界环境与发展委员会《布伦特兰报告：我们共同的未来》	地理学
An Analysis of James E. Lovelock's *Gaia: A New Look at Life on Earth*	解析詹姆斯·E. 拉伍洛克《盖娅：地球生命的新视野》	地理学
An Analysis of Paul Kennedy's *The Rise and Fall of the Great Powers: Economic Change and Military Conflict from 1500—2000*	解析保罗·肯尼迪《大国的兴衰：1500—2000 年的经济变革与军事冲突》	历史
An Analysis of Janet L. Abu-Lughod's *Before European Hegemony: The World System A. D. 1250—1350*	解析珍妮特·L. 阿布-卢格霍德《欧洲霸权之前：1250—1350 年的世界体系》	历史
An Analysis of Alfred W. Crosby's *The Columbian Exchange: Biological and Cultural Consequences of 1492*	解析艾尔弗雷德·W. 克罗斯比《哥伦布大交换：1492 年以后的生物影响和文化冲击》	历史
An Analysis of Tony Judt's *Postwar: A History of Europe since 1945*	解析托尼·朱特《战后欧洲史》	历史
An Analysis of Richard J. Evans's *In Defence of History*	解析理查德·J. 艾文斯《捍卫历史》	历史
An Analysis of Eric Hobsbawm's *The Age of Revolution: Europe 1789–1848*	解析艾瑞克·霍布斯鲍姆《革命的年代：欧洲 1789—1848 年》	历史

An Analysis of Roland Barthes's *Mythologies*	解析罗兰·巴特《神话学》	文学与批判理论
An Analysis of Simon de Beauvoir's *The Second Sex*	解析西蒙娜·德·波伏娃《第二性》	文学与批判理论
An Analysis of Edward W. Said's *Orientalism*	解析爱德华·W. 萨义德《东方主义》	文学与批判理论
An Analysis of Virginia Woolf's *A Room of One's Own*	解析弗吉尼亚·伍尔芙《一间自己的房间》	文学与批判理论
An Analysis of Judith Butler's *Gender Trouble*	解析朱迪斯·巴特勒《性别麻烦》	文学与批判理论
An Analysis of Ferdinand de Saussure's *Course in General Linguistics*	解析费尔迪南·德·索绪尔《普通语言学教程》	文学与批判理论
An Analysis of Susan Sontag's *On Photography*	解析苏珊·桑塔格《论摄影》	文学与批判理论
An Analysis of Walter Benjamin's *The Work of Art in the Age of Mechanical Reproduction*	解析瓦尔特·本雅明《机械复制时代的艺术作品》	文学与批判理论
An Analysis of W.E.B. Du Bois's *The Souls of Black Folk*	解析 W.E.B. 杜波依斯《黑人的灵魂》	文学与批判理论
An Analysis of Plato's *The Republic*	解析柏拉图《理想国》	哲学
An Analysis of Plato's *Symposium*	解析柏拉图《会饮篇》	哲学
An Analysis of Aristotle's *Metaphysics*	解析亚里士多德《形而上学》	哲学
An Analysis of Aristotle's *Nicomachean Ethics*	解析亚里士多德《尼各马可伦理学》	哲学
An Analysis of Immanuel Kant's *Critique of Pure Reason*	解析伊曼努尔·康德《纯粹理性批判》	哲学
An Analysis of Ludwig Wittgenstein's *Philosophical Investigations*	解析路德维希·维特根斯坦《哲学研究》	哲学
An Analysis of G.W.F. Hegel's *Phenomenology of Spirit*	解析 G.W.F. 黑格尔《精神现象学》	哲学
An Analysis of Baruch Spinoza's *Ethics*	解析巴鲁赫·斯宾诺莎《伦理学》	哲学
An Analysis of Hannah Arendt's *The Human Condition*	解析汉娜·阿伦特《人的境况》	哲学
An Analysis of G.E.M. Anscombe's *Modern Moral Philosophy*	解析 G.E.M. 安斯康姆《现代道德哲学》	哲学
An Analysis of David Hume's *An Enquiry Concerning Human Understanding*	解析大卫·休谟《人类理解研究》	哲学

An Analysis of Søren Kierkegaard's *Fear and Trembling*	解析索伦·克尔凯郭尔《恐惧与战栗》	哲学
An Analysis of René Descartes's *Meditations on First Philosophy*	解析勒内·笛卡尔《第一哲学沉思录》	哲学
An Analysis of Friedrich Nietzsche's *On the Genealogy of Morality*	解析弗里德里希·尼采《论道德的谱系》	哲学
An Analysis of Gilbert Ryle's *The Concept of Mind*	解析吉尔伯特·赖尔《心的概念》	哲学
An Analysis of Thomas Kuhn's *The Structure of Scientific Revolutions*	解析托马斯·库恩《科学革命的结构》	哲学
An Analysis of John Stuart Mill's *Utilitarianism*	解析约翰·斯图亚特·穆勒《功利主义》	哲学
An Analysis of Aristotle's *Politics*	解析亚里士多德《政治学》	政治学
An Analysis of Niccolò Machiavelli's *The Prince*	解析尼科洛·马基雅维利《君主论》	政治学
An Analysis of Karl Marx's *Capital*	解析卡尔·马克思《资本论》	政治学
An Analysis of Benedict Anderson's *Imagined Communities*	解析本尼迪克特·安德森《想象的共同体》	政治学
An Analysis of Samuel P. Huntington's *The Clash of Civilizations and the Remaking of World Order*	解析塞缪尔·P. 亨廷顿《文明的冲突与世界秩序重建》	政治学
An Analysis of Alexis de Tocqueville's *Democracy in America*	解析阿列克西·德·托克维尔《论美国的民主》	政治学
An Analysis of John A. Hobson's *Imperialism: A Study*	解析约翰·A. 霍布森《帝国主义》	政治学
An Analysis of Thomas Paine's *Common Sense*	解析托马斯·潘恩《常识》	政治学
An Analysis of John Rawls's *A Theory of Justice*	解析约翰·罗尔斯《正义论》	政治学
An Analysis of Francis Fukuyama's *The End of History and the Last Man*	解析弗朗西斯·福山《历史的终结与最后的人》	政治学
An Analysis of John Locke's *Two Treatises of Government*	解析约翰·洛克《政府论》	政治学
An Analysis of Sun Tzu's *The Art of War*	解析孙武《孙子兵法》	政治学
An Analysis of Henry Kissinger's *World Order: Reflections on the Character of Nations and the Course of History*	解析亨利·基辛格《世界秩序》	政治学
An Analysis of Jean-Jacques Rousseau's *The Social Contract*	解析让-雅克·卢梭《社会契约论》	政治学

An Analysis of Odd Arne Westad's *The Global Cold War: Third World Interventions and the Making of Our Times*	解析文安立《全球冷战：美苏对第三世界的干涉与当代世界的形成》	政治学
An Analysis of Sigmund Freud's *The Interpretation of Dreams*	解析西格蒙德·弗洛伊德《梦的解析》	心理学
An Analysis of William James' *The Principles of Psychology*	解析威廉·詹姆斯《心理学原理》	心理学
An Analysis of Philip Zimbardo's *The Lucifer Effect*	解析菲利普·津巴多《路西法效应》	心理学
An Analysis of Leon Festinger's *A Theory of Cognitive Dissonance*	解析利昂·费斯汀格《认知失调论》	心理学
An Analysis of Richard H. Thaler & Cass R. Sunstein's *Nudge: Improving Decisions about Health, Wealth, and Happiness*	解析理查德·H.泰勒/卡斯·R.桑斯坦《助推：如何做出有关健康、财富和幸福的更优决策》	心理学
An Analysis of Gordon Allport's *The Nature of Prejudice*	解析高尔登·奥尔波特《偏见的本质》	心理学
An Analysis of Steven Pinker's *The Better Angels of Our Nature: Why Violence Has Declined*	解析斯蒂芬·平克《人性中的善良天使：暴力为什么会减少》	心理学
An Analysis of Stanley Milgram's *Obedience to Authority*	解析斯坦利·米尔格拉姆《对权威的服从》	心理学
An Analysis of Betty Friedan's *The Feminine Mystique*	解析贝蒂·弗里丹《女性的奥秘》	心理学
An Analysis of David Riesman's *The Lonely Crowd: A Study of the Changing American Character*	解析大卫·理斯曼《孤独的人群：美国人社会性格演变之研究》	社会学
An Analysis of Franz Boas's *Race, Language and Culture*	解析弗朗兹·博厄斯《种族、语言与文化》	社会学
An Analysis of Pierre Bourdieu's *Outline of a Theory of Practice*	解析皮埃尔·布尔迪厄《实践理论大纲》	社会学
An Analysis of Max Weber's *The Protestant Ethic and the Spirit of Capitalism*	解析马克斯·韦伯《新教伦理与资本主义精神》	社会学
An Analysis of Jane Jacobs's *The Death and Life of Great American Cities*	解析简·雅各布斯《美国大城市的死与生》	社会学
An Analysis of C. Wright Mills's *The Sociological Imagination*	解析C.赖特·米尔斯《社会学的想象力》	社会学
An Analysis of Robert E. Lucas Jr.'s *Why Doesn't Capital Flow from Rich to Poor Countries?*	解析小罗伯特·E.卢卡斯《为何资本不从富国流向穷国？》	社会学

An Analysis of Émile Durkheim's *On Suicide*	解析埃米尔·迪尔凯姆《自杀论》	社会学
An Analysis of Eric Hoffer's *The True Believer: Thoughts on the Nature of Mass Movements*	解析埃里克·霍弗《狂热分子：群众运动圣经》	社会学
An Analysis of Jared M. Diamond's *Collapse: How Societies Choose to Fail or Survive*	解析贾雷德·M.戴蒙德《大崩溃：社会如何选择兴亡》	社会学
An Analysis of Michel Foucault's *The History of Sexuality Vol. 1: The Will to Knowledge*	解析米歇尔·福柯《性史（第一卷）：求知意志》	社会学
An Analysis of Michel Foucault's *Discipline and Punish*	解析米歇尔·福柯《规训与惩罚》	社会学
An Analysis of Richard Dawkins's *The Selfish Gene*	解析理查德·道金斯《自私的基因》	社会学
An Analysis of Antonio Gramsci's *Prison Notebooks*	解析安东尼奥·葛兰西《狱中札记》	社会学
An Analysis of Augustine's *Confessions*	解析奥古斯丁《忏悔录》	神学
An Analysis of C.S. Lewis's *The Abolition of Man*	解析 C.S. 路易斯《人之废》	神学

图书在版编目（CIP）数据

解析约翰·斯图亚特·穆勒《功利主义》：汉、英 / 帕特里克·汤姆（Patrick Tom），桑德尔·韦尔克霍芬（Sander Werkhoven）著；陈琦译.
—上海：上海外语教育出版社，2019
（世界思想宝库钥匙丛书）
ISBN 978-7-5446-5956-7

Ⅰ.①解… Ⅱ.①帕…②桑…③陈… Ⅲ.①穆勒（Mill, John Stuart 1806—1873）-功利主义—研究—汉、英 Ⅳ.①B561.42②B82-064

中国版本图书馆CIP数据核字（2019）第152285号

This Chinese-English bilingual edition of *An Analysis of John Stuart Mill's* Utilitarianism is published by arrangement with MACAT International Limited.
Licensed for sale throughout the world.

本书汉英双语版由Macat国际有限公司授权上海外语教育出版社有限公司出版。
供在全世界范围内发行、销售。

图字：09 – 2018 – 549

出版发行：**上海外语教育出版社**
（上海外国语大学内）　邮编：200083
电　　话：021-65425300（总机）
电子邮箱：bookinfo@sflep.com.cn
网　　址：http://www.sflep.com
责任编辑：田慧肖

印　　刷：上海宝山译文印刷厂
开　　本：890×1240　1/32　印张 5.5　字数 114千字
版　　次：2019 年 9 月第 1 版　2019 年 9 月第 1 次印刷
印　　数：2 100 册

书　　号：ISBN 978-7-5446-5956-7
定　　价：30.00 元

本版图书如有印装质量问题，可向本社调换
质量服务热线：4008-213-263　电子邮箱：**editorial@sflep.com**